New Middle and Late Anisian (Middle Triassic) ammonoid faunas from northwestern Nevada (USA): taxonomy and biochronology

by

Claude Monnet and Hugo Bucher

Acknowledgement

Financial support for the publication of this number of Fossils and Strata was provided by the Swiss National Science Foundation.

Contents

New Middle and Late Anisian (Middle Triassic) ammonoid faunas from northwestern Nevada (USA): taxonomy and biochronology

CLAUDE MONNET & HUGO BUCHER

Monnet, C. & Bucher, H. **2005 12 01**: New Middle and Late Anisian (Middle Triassic) ammonoid faunas from northwestern Nevada (USA): taxonomy and biochronology. *Fossils and Strata*, No. 52, pp. 1–121. Oslo. ISSN 0300-9491.

An intensive investigation of the Fossil Hill Member (northwestern Nevada, United States) leads to the recognition of new ammonoid faunas that bracket the Middle/Late Anisian boundary. These faunas are of great significance for biochronological correlation across the low palaeolatitude belt from the eastern Pacific to the western end of the Tethys. Three new sections in the Augusta Mountains have yielded a rich faunal succession, whose upper part correlates with the resampled lower portion of the classic Fossil Hill section in the Humboldt Range. In the new biostratigraphical sequence, the scope of the latest middle Anisian *Shoshonensis* Zone is expanded by the introduction of a new uppermost subdivision, namely the *Bulogites mojsvari* Subzone, which correlates with the Grossreifling fauna of the western Tethys. The *Gymnotoceras weitschati* Zone and *Gymnotoceras mimetus* Zone are introduced at the base of the Late Anisian, in ascending order. The *Weitschati* Zone, composed of the *Billingsites cordeyi* and *Rieberites transiformis* subzones, is approximately equivalent to the ill-defined *Trinodosus* Zone of the Tethyan realm. Included in the *Mimetus* Zone are the *Dixiceras lawsoni* and *Marcouxites spinifer* subzones. The *Gymnotoceras rotelliformis* Zone, which was formerly considered as a correlative of the *Trinodosus* Zone, was previously subdivided into five subzones, mainly on the basis of various species referred to as "*Paraceratites*". These species, whose respective ranges are shown largely to overlap, are reassigned to the new genera *Silberlingia*, *Ceccaceras*, and *Brackites*. Hence, the number of subdivisions of the *Rotelliformis* Zone is reduced to two, namely the *Brackites vogdesi* and *Gymnotoceras blakei* subzones, in ascending order. Eleven genera (i.e. *Chiratites*, *Billingsites*, *Dixiceras*, *Jenksites*, *Rieppelites*, *Rieberites*, *Marcouxites*, *Silberlingia*, *Ceccaceras*, *Brackites*, *Oxylongobardites*) and 15 species (i.e. *Balatonites hexatuberculatus*, *Chiratites retrospinosus*, *C. bituberculatus*, *Billingsites cordeyi*, *B. escargueli*, *Gymnotoceras weitschati*, *G. mimetus*, *Jenksites flexicostatus*, *Rieppelites boletzkyi*, *R. shevyrevi*, *Rieberites transiformis*, *Silberlingia praecursor*, *Ceccaceras stecki*, *Brackites spinosus*, *Oxylongobardites acutus*) are newly described. "*Ceratites*" *lawsoni* and "*Ceratites*" *spinifer*, as previously described by Smith (1914) and subsequently synonymized by Silberling & Nichols (1982), are recognized as valid species and are assigned to the new genera *Dixieceras* and *Marcouxites*, respectively.

Key words: Ammonoids; Anisian (Middle Triassic); Fossil Hill Member; northwestern Nevada (USA); biostratigraphy; taxonomy.

Claude Monnet [claude.monnet@pim.unizh.ch] & Hugo Bucher [hugo.fr.bucher@pim.unizh.ch], Paläontologisches Institut und Museum, Universität Zürich, Karl Schmid Strasse 4, CH-8006 Zürich, Switzerland.

Introduction

During the past four decades, the ammonoid biochronology of the North American Triassic has been significantly refined, thanks largely to a wealth of successive fossiliferous levels that occur in Nevada and Canada. Triassic ammonoids from Nevada were first made known by the pioneer contributions of Hyatt & Smith (1905) and

Smith (1914), who described, in part, the Anisian faunas from the classic Fossil Hill locality, as well as other sites in the northern Humboldt Range. High-resolution data were first made available by Silberling & Nichols (1982) who established the main biostratigraphical ammonoid succession for the Anisian, with special emphasis on the Late Anisian. Their work, based on bed-by-bed sampling in the Fossil Hill–Saurian Hill area (Humboldt Range), highlighted what is considered to be the world's most complete sequence of low-palaeolatitude, Late Anisian ammonoid faunas. Hence, the Humboldt Range, and especially the Fossil Hill area, has become a standard for Late Anisian ammonoid succession (Silberling & Tozer, 1968).

Subsequently, Bucher (1988, 1989, 1992a, b, 1994) improved upon the understanding of the Fossil Hill succession and enlarged the taxonomic and biostratigraphical scope of the Early and Middle Anisian. For the Middle Anisian succession, the Augusta Mountains yielded the most significant sections. Recent investigations in this range led to the discovery of new faunas straddling the Middle/Late Anisian boundary, and the present contribution focuses on the taxonomy and biostratigraphy of these faunas. Their implication for low-palaeolatitude biochronology, and correlation between the eastern Pacific and the Tethys realm, as well as their palaeobiogeographical and phylogenetic implication, will be dealt with elsewhere.

Geological context

All of the studied ammonoid faunas occur in the Fossil Hill Member, which is common to both the Prida and Favret Formations of the Star Peak Group. Silberling & Wallace (1969) and Nichols & Silberling (1977) studied the stratigraphy of the Star Peak Basin. In terms of lithology, the Fossil Hill Member is fairly uniform and consists essentially of silty shales alternating with dark, micritic, often laminated, either lenticular or thin-bedded limestone, with a relatively high organic matter content. These rocks were deposited below storm wave base, in a euxinic environment, and typically yield planktonic, pseudo-planktonic, and nektonic organisms (radiolarians, ostracods, conodonts, halobiid bivalves, cephalopods). Ammonoids are relatively well preserved, with a calcified external shell.

The Fossil Hill Member covered most, if not all, of the Triassic Star Peak Basin, which was a plate-bound, complex carbonate platform (Fig. 1A) largely controlled by differential uplift and subsidence (Nichols & Silberling 1977) resulting from an incipient extension of northwestern Nevada during Anisian time (Wyld 2000). Accordingly, the effects of synsedimentary tectonics are expressed by the significant lateral variations in age of the lower and upper boundaries of the Fossil Hill Member. Since the greatest lateral extension of the Fossil Hill Member occurred during the Late Anisian, the number of well-exposed sections straddling the Middle/Upper Anisian boundary is comparatively few. Among these, the

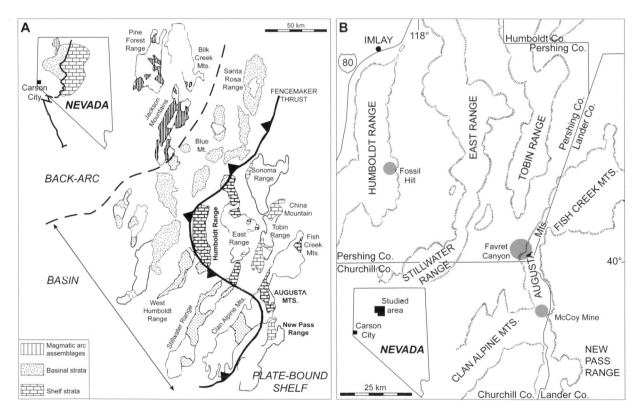

Fig. 1.   A: Simplified geological map of northwestern Nevada (after Wyld 2000, modified). B: Location map of the sampled areas (Humboldt Range, Augusta Mountains, and New Pass Range).

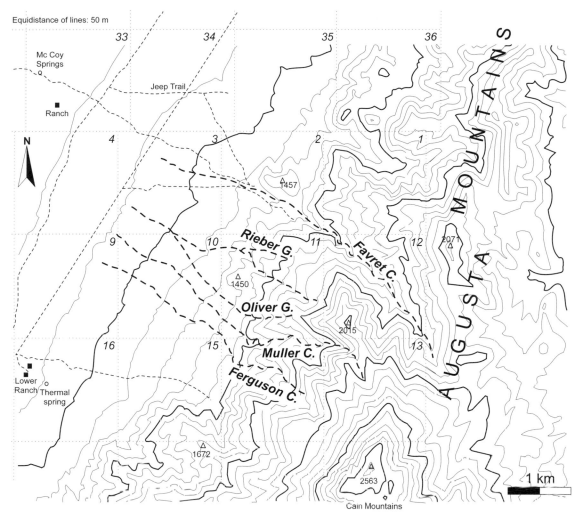

Fig. 2. Location map of the sampled sections in the Augusta Mountains (Oliver Gulch, Ferguson Canyon, Rieber Gulch, Muller Canyon, and Favret Canyon).

best sections (Oliver Gulch, Rieber Gulch, and Ferguson Canyon) are found on the western slopes of the Augusta Mountains (Pl. 1), south of Favret Canyon (Figs. 1B, 2). The sampled sections yielded numerous fossiliferous levels distributed throughout up to 100 m of strata (Figs. 3–7). The presence of marker beds (MB1 to MB6; Figs. 3–7) enabled the construction of a composite section for this area.

Additional samples were also obtained from the McCoy Mine area in the New Pass Range (Fig. 1B), but the comparatively poor exposures in that area did not permit the thorough sampling that was possible in the Augusta Mountains. Finally, the Fossil Hill section in the Humboldt Range (Fig. 1B) was resampled within the scope of this study, in order to link this classic section with the new faunal sequence from the Augusta Mountains, and to facilitate construction of a comprehensive composite faunal succession. Middle/Late Anisian transition beds also occur in the southern Tobin Range, but

they have been largely offset by extensive faulting of Tertiary age.

Biostratigraphy

Silberling & Tozer (1968), Silberling & Wallace (1969), and Silberling & Nichols (1982) established the basic biostratigraphical succession for the Anisian ammonoid faunas of Nevada. Then, Bucher (1988, 1989, 1992a, b, 1994) focused on the Early and Middle Anisian part of the Fossil Hill record, and subsequently developed a comprehensive ammonoid biostratigraphical zonation for these substages. Our investigations focus on the transition between the Middle and Late Anisian record, i.e. the interval bracketed by the *Shoshonensis* and *Rotelliformis* zones (Fig. 8), which have been described by Bucher (1992b) and Silberling & Nichols (1982), respectively.

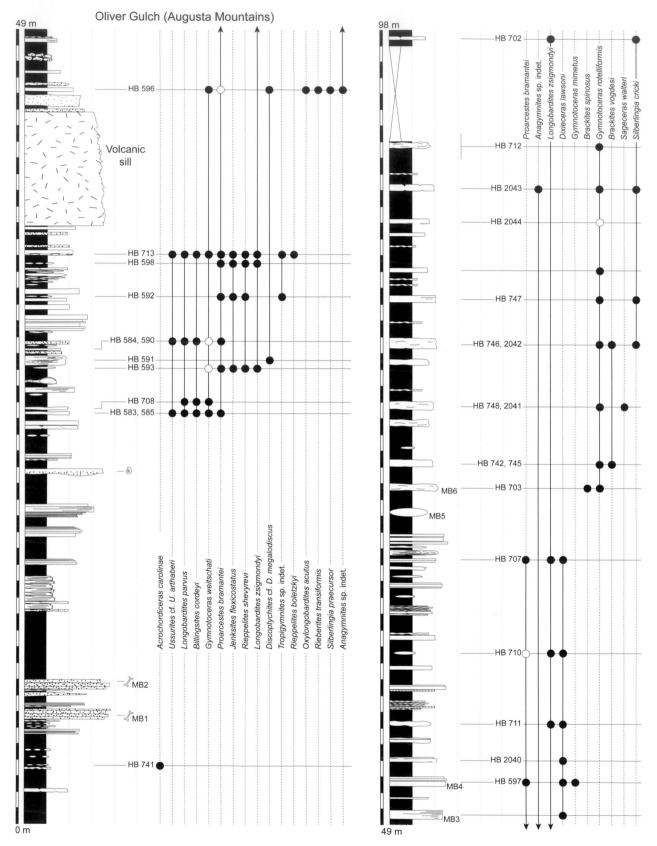

Fig. 3.   Distribution of ammonoid taxa in the Oliver Gulch section. MB: marker beds. Open dots indicate occurrences based only on fragmentary or poorly preserved material.

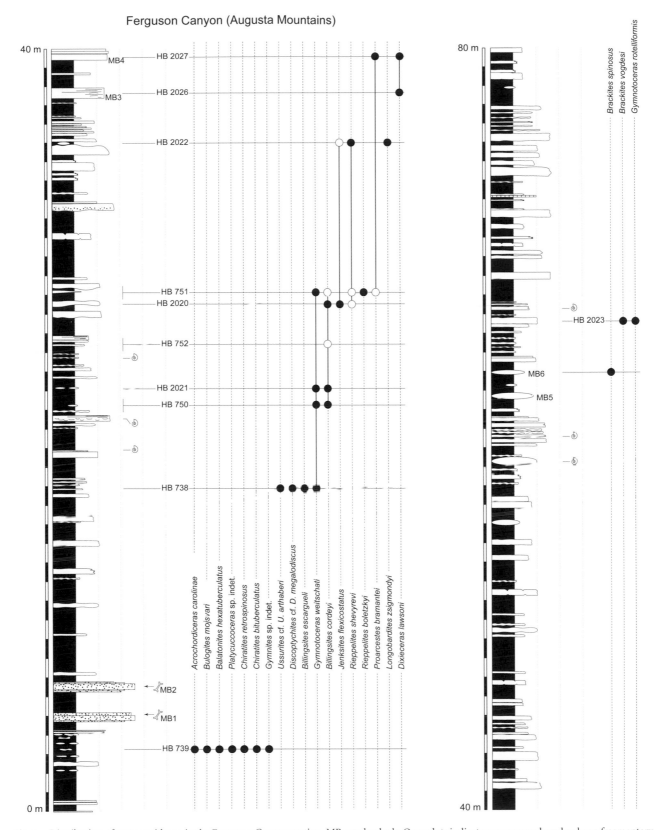

Fig. 1.　Distribution of ammonoid taxa in the Ferguson Canyon section. MB: marker beds. Open dots indicate occurrences based only on fragmentary or poorly preserved material.

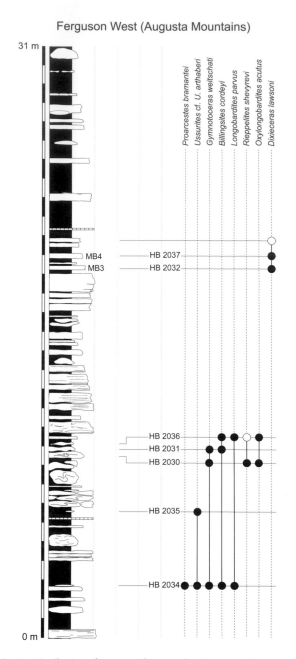

Fig. 5. Distribution of ammonoid taxa in the Ferguson West section. MB: marker beds. Open dots indicate occurrences based only on fragmentary or poorly preserved material.

Our biostratigraphical data set includes 42 species (Fig. 9) from five sections (Oliver Gulch, Rieber Gulch, Ferguson Canyon, Ferguson West, Fossil Hill) and several isolated samples (e.g. McCoy Mine; Fig. 10). This data set is analysed by means of the Unitary Associations method (Guex 1991). The resulting zonation is composed of discrete association zones, called UA zones (closely allied to Oppel zones and concurrent range zones), which are maximal sets of intersecting ranges of taxa and are the finest subdivisions based on the association concept.

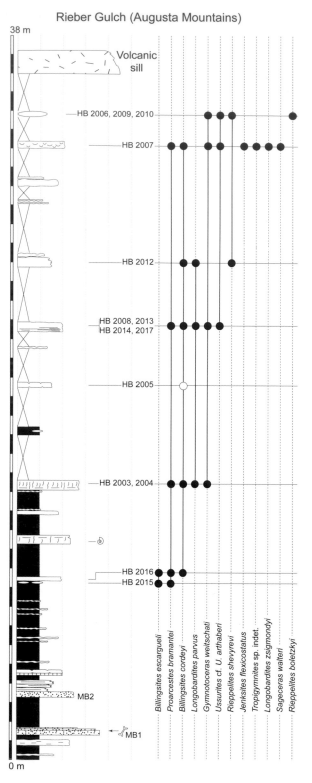

Fig. 6. Distribution of ammonoid taxa in the Rieber Gulch section. MB: marker beds. Open dots indicate occurrences based only on fragmentary or poorly preserved material.

Compared with empirical association zones and concurrent range zones, this method leads to substantially higher resolved zonations, even for ammonoids (see Monnet &

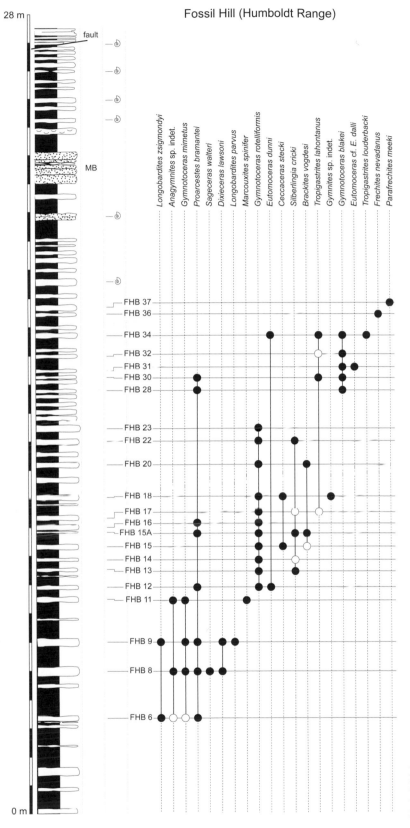

Fossil Hill (Humboldt Range)

Fig. 7. Distribution of ammonoid taxa in the lower part of the Fossil Hill section. MB: brown calcareous sandstone unit of Silberling & Nichols (1982). Open dots indicate occurrences based only on fragmentary or poorly preserved material.

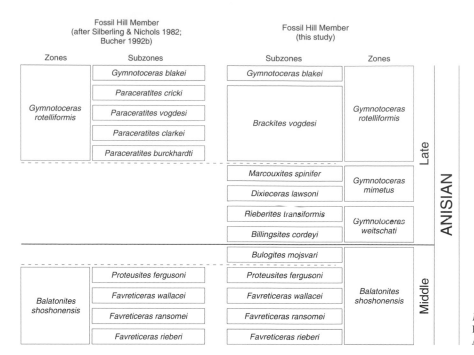

Fig. 8.   Revised ammonoid zonation of the Fossil Hill Member around the Middle/Late Anisian boundary.

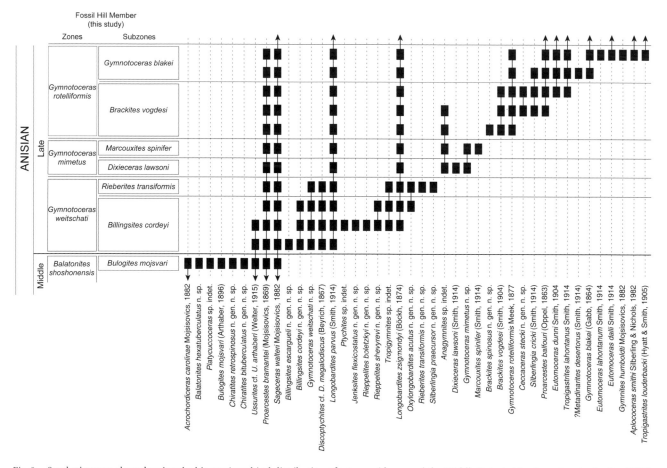

Fig. 9.   Synthetic range chart showing the biostratigraphical distribution of ammonoids around the Middle/Late Anisian boundary in the Fossil Hill Member.

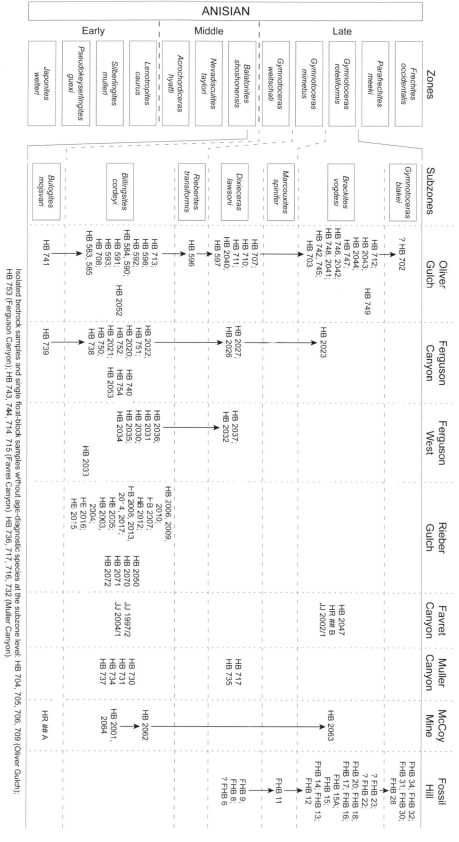

Fig. 10. Correlations and superpositional relationships of bedrock-controlled samples and single isolated blocks in the studied areas (Augusta Mountains, New Pass Range, Humboldt Range). Question marks indicate samples in sequence but without age-diagnostic ammonoids.

Bucher 2002), which are traditionally acknowledged as one of the most useful groups for dating Mesozoic marine rocks. For an extensive methodological description of the Unitary Associations, the reader is referred to Guex (1991).

Biochronological processing yielded 12 Unitary Associations spanning the *Shoshonensis–Rotelliformis* time interval, and the faunal content of each Unitary Association is shown in Figure 9. Reunions of Unitary Associations (i.e. UA zones) were given a subzone rank, thus leading to the establishment of seven subzones. These subzones were then grouped into zones on the basis of the similarity of their faunal content.

These results make possible the greatly expanded Middle and Late Anisian ammonoid zonation for the Fossil Hill Member shown in Figure 8. The occurrence of distinct ammonoid assemblages bracketed by the *Shoshonensis* and *Rotelliformis* zones enables the recognition of two new Late Anisian zones below the *Rotelliformis* Zone. Each new zone contains two subzones (Fig. 8). *Gymnotoceras weitschati* n. sp. and *Gymnotoceras mimetus* n. sp. are the index species for the older and younger zone, respectively. The scope of the *Shoshonensis* Zone is also expanded with the introduction of the *Bulogites mojsvari* Subzone as its uppermost subdivision. Finally, the *Rotelliformis* Zone itself shows important changes because the first four subzones defined by Silberling & Nichols (1982) are merged into a single subzone called the *Brackites vogdesi* Subzone. As adopted here, the Middle/Late Anisian boundary is marked by the disappearance of the genera *Balatonites*, *Platycuccoceras*, *Acrochordiceras*, and *Bulogites*, as well as by the appearance of the genera *Billingsites*, *Rieppelites*, *Jenksites*, and *Longobardites*. The boundary is also marked by a change at the species level within the genus *Gymnotoceras*.

Shoshonensis Zone

Index species. – *Balatonites shoshonensis* Hyatt & Smith, 1905

Type locality. – USGS Mesozoic localities M501 and M635, Wildhorse mining district (northern New Pass Range). This zone is also well documented in Favret Canyon (Augusta Mountains).

Originally defined by Silberling & Tozer (1968), the scope of this zone was subsequently expanded to four distinct subzones (*Rieberi*, *Ransomei*, *Wallacei*, and *Fergusoni*, in ascending order) by Bucher (1992b), to which the reader is referred for a complete description. The new faunas described here allow the recognition of a distinct, new subzone, the *Mojsvari* Subzone, as the uppermost subdivision of the zone.

Mojsvari Subzone

Index species. – *Bulogites mojsvari* (Arthaber, 1896)

Type locality. – Loc. HB 739, Ferguson Canyon (Augusta Mountains)

Other occurrences. – This subzone is also documented in the Oliver Gulch section (Augusta Mountains) and in the McCoy Mine area (New Pass Range), as indicated in Fig. 10.

In the Augusta Mountains, this fauna typically occurs in small-sized (dm), early diagenetic calcareous nodules embedded in a predominantly shaly and recessive interval. Our best collection comes from the base of the Ferguson Canyon section (loc. HB 739), but it is also documented from the same horizon in the Oliver Gulch section (loc. HB 741). H. Rieber (Zürich) also collected a float specimen of *Bulogites mojsvari* in the McCoy Mine area (Pl. 4: 3), thus confirming the existence of this subzone in the New Pass Range as well. *Bulogites mojsvari* is designated as the index species because it is a very short-ranging species, first described by Arthaber (1896) from Grossreifling (Austria). *Chiratites* n. gen. and *Balatonites hexatuberculatus* n. sp. are presently known to occur exclusively in the *Mojsvari* Subzone. *Acrochordiceras carolinae* has its youngest occurrence in this subzone.

Weitschati Zone

Index species. – *Gymnotoceras weitschati* n. sp.

Type locality. – Oliver Gulch (Augusta Mountains)

Other occurrences. – This zone is documented in almost all sections from the Augusta Mountains (except in Favret Canyon) and is also recognized in the McCoy Mine area (New Pass Range), as indicated in Fig. 10.

The index species, *G. weitschati*, is found in almost all samples and sections of the zone, but is particularly abundant in Rieber Gulch. This zone has been subdivided into two subzones, the *Cordeyi* Subzone and the *Transiformis* Subzone, in ascending order.

Cordeyi Subzone

Index species. – *Billingsites cordeyi* n. gen. n. sp.

Type locality. – Loc. HB 713, Oliver Gulch (Augusta Mountains)

Other occurrences. – This subzone is well represented south of Favret Canyon in the Augusta Mountains, and it also occurs in the McCoy Mine area in the New Pass Range (Fig. 10).

It is characterized by the appearance of *Billingsites* n. gen. and by the diagnostic occurrence of *Jenksites* n. gen. and *Rieppelites* n. gen.

Transiformis Subzone

Index species. – *Rieberites transiformis* n. gen. n. sp.

Type locality. – Loc. HB 596, Oliver Gulch (Augusta Mountains).

Other occurrences. – Although this subzone is known only from a single locality (loc. HB 596), its abundant and distinctive faunal association, well bracketed in the sequence, justifies its erection to the subzone level (Fig. 10).

This assemblage is characterized by the restricted occurrence of *Rieberites transiformis* n. gen. n. sp., the disappearance of *Gymnotoceras weitschati* n. sp., and the appearance of *Silberlingia* n. gen.

Mimetus Zone

Index species. – *Gymnotoceras mimetus* n. sp.

Type locality. – Fossil Hill (Humboldt Range)

Other occurrences. – This zone is also well documented in the Augusta Mountains (Fig. 10).

Silberling & Nichols (1982) did not report faunas presently assigned to the *Mimetus* Zone from Fossil Hill. However, *Dixieceras lawsoni*, which is a characteristic element of this zone, was previously described from Fossil Hill by Smith (1914), but it was tentatively synonymized with younger, more or less homeomorphic taxa (such as *Gymnotoceras*) by Silberling & Nichols (1982). The *Mimetus* Zone consists of the *Lawsoni* Subzone and the *Spinifer* Subzone, as newly documented at Fossil Hill. The succession between the *Weitschati*, *Mimetus*, and *Rotelliformis* Zones is clearly established in the Oliver Gulch section.

Lawsoni Subzone

Index species. – *Dixieceras lawsoni* (Smith, 1914)

Type locality. – Loc. FHB 9, Fossil Hill (Humboldt Range)

Other occurrences. – Abundant faunas of this subzone are also documented at Saurian Hill (Humboldt Range), Oliver Gulch, Ferguson Canyon, and Muller Canyon (Augusta Mountains), as indicated in Fig. 10.

This subzone is characterized by the restricted occurrence of *"Ceratites" (Philippites) lawsoni* Smith, 1914,

which we consider to be a valid species. It is morphologically distinct from the younger *Gymnotoceras rotelliformis*, with which it was synonymized by Spath (1934), and from *Frechites* sp., with which it was synonymized by Silberling & Nichols (1982). It is herein assigned to the new genus *Dixieceras*.

Spinifer Subzone

Index species. – *Marcouxites spinifer* (Smith, 1914)

Type locality. – Loc. FHB 11, Fossil Hill (Humboldt Range)

This subzone is presently known only from Fossil Hill (FHB 11). *"Ceratites" spinifer* was erected by Smith (1914), and subsequently synonymized with the younger *Frechites nevadanus* by Silberling & Nichols (1982). However, *"Ceratites" spinifer* Smith, 1914 is considered as a valid species and is assigned to the new genus *Marcouxites*.

Rotelliformis Zone

Index species. – *Gymnotoceras rotelliformis* Meek, 1877

Type locality. – Fossil Hill (Humboldt Range)

Other occurrences – This zone occurs in almost all sections of the Humboldt Range, the Augusta Mountains, and the New Pass Range (Fig. 10), as well as in the northern East Range, southern Tobin Range, and China Mountain (see Nichols & Silberling 1977).

As established by Silberling & Nichols (1982), the *Rotelliformis* Zone includes five subzones, which are essentially based on the succession of four supposedly non-overlapping species of *"Paraceratites"* (*"P."* burckhardti, *"P."* clarkei, *"P."* vogdesi, and *"P."* cricki), followed by the beyrichitin *Gymnotoceras blakei*. New data obtained from Fossil Hill and the Augusta Mountains indicate that the four older subzones of these authors correspond in fact to a single subzone, herein designated as the *Vogdesi* Subzone. This simplification is based on the two following arguments. First, *"P."* burckhardti and *"P."* cricki are demonstrated to be morphological variants of a single species. Second, the ranges of the three remaining valid species of *"Paraceratites"* are actually found to overlap, both in the Fossil Hill type area and in the Augusta Mountains. The younger *Blakei* Subzone, as defined by Silberling & Nichols (1982), remains unchanged.

Vogdesi Subzone

Index species. – *Brackites vogdesi* (Smith, 1904)

Type locality. – Loc. FHB 15A, Fossil Hill (Humboldt Range)

Other occurrences. – See Figure 10.

The index species *Brackites vogdesi* is the more easily recognizable species of this subzone. *Ceccaceras stecki* n. gen. n. sp. is found exclusively in this subzone but is relatively rare, whereas *Silberlingia cricki* (Smith, 1914) extends into the younger *Blakei* Subzone. *Eutomoceras* first appears in this subzone, thus making its earliest appearance older than previously documented by Silberling & Nichols (1982).

Systematics

Intraspecific variability

It is worth noting that, when available, sufficiently large samples allow the recognition of the First Buckman's Law of Covariation within assemblages obtained from single beds. Hence, most of the species described hereafter possess a continuous intraspecific variation ranging from "slender" forms, which are relatively involute, compressed, and weakly ornamented, to "robust" forms, which are more evolute, more depressed, and with coarser ornamentation. Within a single species, the frequency of these variants is displayed by a typical normal (Gaussian) bell-shaped curve. This concept of intraspecific variation has been well demonstrated by Westermann (1966), Silberling & Nichols (1982), and Dagys & Weitschat (1993), among others.

It is also worth noting that most of the diagnostic characteristics of a species are better expressed by the robust variants. Slender variants are less well discernible and tend to converge across closely related species or even genera, thus making recognition of intraspecific variation a crucial characteristic for correct identification. Converging morphologies of slender variants belonging to different species illustrate such difficulties, and are especially common among beyrichitins, for example.

Some of the Nevada species described in this study are very similar to certain Alpine species. Despite their priority, the validity of many of the Alpine taxa remains highly conjectural because their respective range of intraspecific variation and ontogenetic development are poorly known, and data regarding their superpositional relationships are exceedingly scarce. Essentially, Alpine ammonoids were differentiated from each other utilizing the typological approach, which is hardly compatible with the population approach used for the Nevada faunas. Ultimately, a taxonomic standardization between the Alpine and Nevada faunas will be required, but it should be based on new and accurate Alpine documentation, which is beyond the scope of this study. Until suitable Alpine data become available, it is preferable to develop an independent taxonomy for the Nevada faunas in order to avoid propagation of ill-defined taxa, and to preserve and exploit the high potential of the Nevada successions for biochronology and evolutionary patterns of Anisian ammonoid lineages.

Measurements and statistical tests

The quantitative morphological range of each species is expressed utilizing the four classic geometrical parameters of the ammonoid shell: shell diameter (D), whorl height (H), whorl width (W), and umbilical diameter (U).

The variability of the three parameters H, W, and U is expressed graphically and statistically. Data for *Billingsites cordeyi* n. gen. n. sp., for example, are displayed in Figure 11. For each species represented by sufficiently large samples (i.e. at least 30 specimens), parameters are plotted in absolute values (H, W, U) and in percentages against the size-related parameter D (H/D, W/D, U/D) (Fig. 11A, B). The normality of each of the three parameters (H, W, U) is statistically tested numerically by means of a Lilliefors test (Fig. 11C, E, G), and graphically by means of a probability plot (Fig. 11D, F, H). The Lilliefors test evaluates the hypothesis that data X has a normal distribution with unspecified mean and variance, against the alternative that X does not have a normal distribution. This test compares the empirical distribution of X with a normal distribution having the same mean and variance as X. In Figure 11C, E, G, the result of the test is indicated in the legend of the calculated normal curve associated with X: the label "Normal" indicates that the test cannot reject the hypothesis that the distribution of X is normal (with a confidence level of 95%); the label "Not Normal" indicates that the hypothesis of X having a normal distribution should be rejected at the 5% confidence level. Additionally, the normal probability plot allows graphical testing of normality (Fig. 11D, F, H). This type of graph plots sampled data (with the symbol "+"). Superimposed on the plot is a robust linear fit of the sample order statistics. If the data are derived from a normal distribution, the plot will appear linear. Other probability density functions will generate departure from a linear plot.

Therefore, the results displayed in Figure 11 indicate that the three parameters H, W, and U have a continuous range and relatively narrowly fluctuating values throughout growth. Umbilical diameter (U) has the highest variability. In percentages against the size-related parameter D, the three parameters appear more variable, with a clear change throughout growth, at least for H and U in the early whorls of the species. From a statistical point of view, H and U do not have a normal distribution, thus

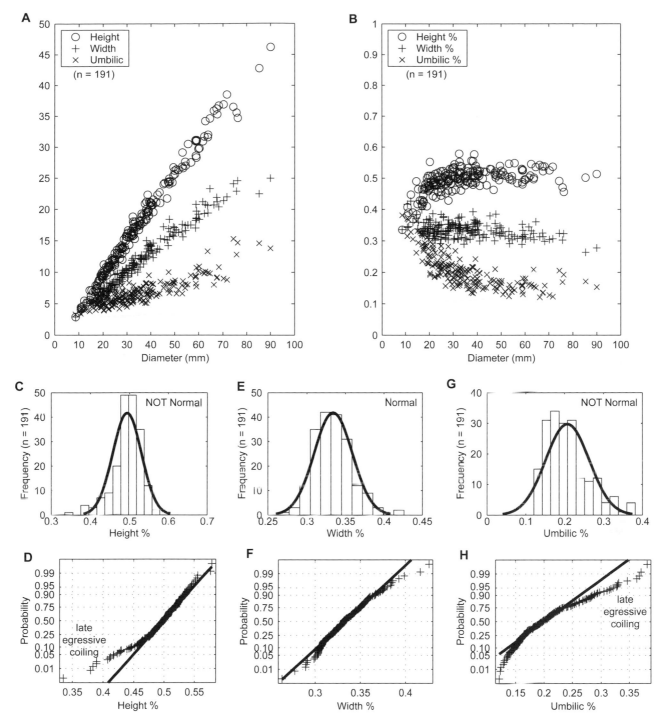

Fig. 11. Biometric analysis of *Billingsites cordeyi* n. gen. n. sp. from the Augusta Mountains (*Cordeyi* Subzone, *Weitschati* Zone, Late Anisian). A: Scatter diagram of H, W, and U against corresponding diameter. B: Scatter diagram of H/D, W/D, and U/D against corresponding diameter. C, E, G: Histograms illustrating the continuous intraspecific variability of *B. cordeyi* for H, W, and U, respectively; the label indicates if the distribution is statistically normal (see text for explanations). D, F, H: probability plots of H, W, and U, respectively, which help to graphically assess the normality of each parameter. In this case, *B. cordeyi* shows a departure from normality towards lower whorl heights and towards larger umbilical diameters, which in fact reflects the egressive coiling of submature specimens.

differing from W, which has a definite normal distribution. Assuming that all parameters should have a normal distribution, it indicates that a departure from normality may result from a data entry error, a poor measurement,

or a change in the system that generated the data. In other words, departures from normality, such as those displayed by H and U for the species *Billingsites cordeyi*, may result from a change in the relative proportions of the ammonoid shell through ontogeny. In fact, the graphs highlight that a significant number of specimens tend to have a smaller whorl height and a larger umbilical diameter. Examination of the morphology of the species for each ontogenetic stage leads to the interpretation of these data as resulting from definite egressive coiling of the shell in the outer whorls (see Pls. 6, 7). It is noteworthy that, after removing the near-adult specimens of this species from the database, the distribution of all three parameters, H, W, and U, appears to be normal. Hence, the Lilliefors test and the normal probability plot also help to quantitatively highlight the ontogenetic changes of the studied species.

A quick review of all species indicates that each species, if represented by sufficiently large samples, appears to have normal (Gaussian) distribution of its specimens for all three parameters (H, W, U). Departures from normal distribution for some species result from ontogenetic changes, such as egressive coiling during the adult stage of a species or a marked change in whorl proportions during juvenile stages. It is also worth noting that the umbilical diameter (U) is usually more variable than the two other parameters.

Finally, all studied species of several subfamilies are quantitatively compared by means of box and mean plots. The box plot displays notched boxes, which represent the range covered by about 99% of the specimens from a normally distributed sample. Boxes are bounded by the 25th and 75th percentiles and bisected by the 50th percentile (median). Hence, box plots allow one to determine whether data from several groups (in this case, species) have a common distribution, i.e. to determine whether or not the groups differ by the measured parameters, as indicated by the amount of overlap between the different notched boxes. The mean plot produces the mean and its associated 95% confidence interval for each species. This type of graph allows one to perform a graphical multiple comparison of the mean of each species, and therefore to test which species have a significantly different mean by assessing which mean and its associated confidence interval overlap.

For example, Figure 12 displays the box (Fig. 12A) and mean (Fig. 12B) plots for W (whorl width) for the new genus *Billingsites*, which includes the two species *B. escargueli* and *B. cordeyi*. *B. escargueli* exhibits a whorl width distribution that slightly overlaps the other species. Moreover, its whorl width has a significantly different mean value: therefore, *B. escargueli* is distinguished from the other species by its significantly more depressed whorl section.

Allometry

The graphs and statistical tests presented above focus on the analysis of single parameters, which reflect phenotypical differences. In order to detect and quantify possible

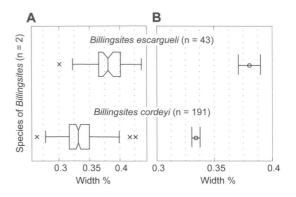

Fig. 12.   Statistical summary of the species belonging to the new genus *Billingsites*. A: Box plot showing the distribution of all values of W for each species. B: Mean plot showing the mean of W and its 95% confidence interval for each species. Here, it highlights that the species *B. escargueli* differs significantly from *B. cordeyi* by a thicker shell. See text for explanations of the plot.

heterochronic processes among the studied species, it is necessary to identify possible changes in size-based allometries of the geometry of the shell dimensional parameters. Therefore, the growth trajectories of H, W, and U are investigated in order to determine allometry with respect to D.

Because isometric growth implies that the parameter has a constant ratio as a function of D, i.e. follows a linear equation, and because allometric growth conforms to an exponential-like equation, the values of each parameter are fitted both by a linear and power equation by means of the reduced major axis fitting method. The two resulting fitted curves are then tested by means of the coefficient of determination, the dispersion of the residuals, and the Z-statistic associated with the allometric exponent. Classically, a better fit is obtained when the correlation coefficient tends towards 1 and when the residuals tend not to be evenly scattered. The Z-statistic tests the null hypothesis that the allometric exponent is equal to 1 (i.e. in fact, isometric growth) at a confidence level of 95%.

For example, Figure 13A–D displays the values of U (umbilical diameter) for *Billingsites cordeyi* with isometric (Fig. 13A, B) and allometric (Fig. 13C, D) fitted curves, their associated residuals and coefficient of determination, and the allometric exponent and its associated Z-statistic. *Billingsites cordeyi* appears to have allometric growth of its umbilical diameter with an allometric coefficient of approximately 0.63. Even if the coefficient of determination of the linear fit is closer to 1, the residuals of fitting and the Z-statistic clearly indicate allometry. Figure 13E–F exhibits the resulting growth curves of the two species of *Billingsites* for whorl width and umbilical diameter. From a quantitative point of view, *B. escargueli* differs from *B. cordeyi* by having isometric growth of its umbilical diameter and by its slightly but always greater whorl width, thus yielding an umbilical growth curve parallel to that of *B. cordeyi*.

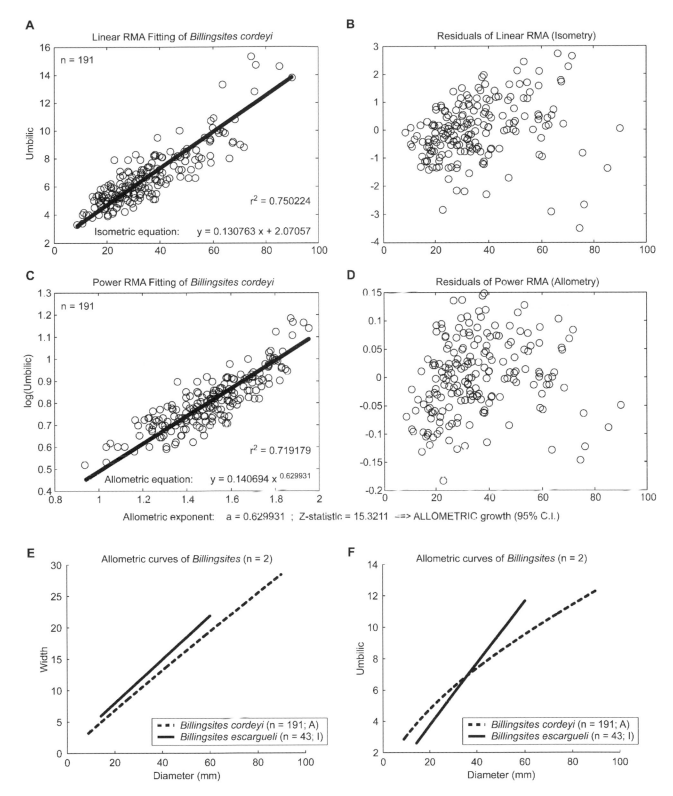

Fig. 13. Examples of allometric growth curves of species belonging to the genus *Billingsites*. A: Test of linear fit (isometry) with its coefficient of determination ($r^2 = 0.75$). B: Residuals of the linear fit. C: Alternative test of power fit (allometry) with its coefficient of determination ($r^2 = 0.72$). D: Residuals of the power fit. The allometric exponent and its Z-statistic are reported below; these allow the evaluation of the significance of the fit; in this case, the umbilical growth of *B. cordeyi* is significantly allometric. E, F: Calculated growth curves of each species of the genus *Billingsites* for the whorl width and the umbilical diameter, respectively. The labels in parentheses indicate whether a growth curve is allometric (A) or isometric (I) according to the calculated Z-statistic. From a systematic point of view, these graphs show that, on the one hand, *B. cordeyi* and *B. escargueli* have a similar growth of width (although *B. escargueli* is always slightly thicker) and, on the other, they have different umbilical growth (for example, *B. escargueli* has a proportionally increasing umbilical diameter with growth).

Systematic descriptions

All systematic descriptions follow the classification of Tozer (1981). Repositories for figured and measured specimens are abbreviated PIMUZ (Paläontologisches Institut und Museum der Universität Zürich) or USNM (National Museum of Natural History, Washington, D.C.).

Occurrences of taxa described herein include the number of specimens obtained from each locality. For example, HB 2030 (12) indicates that twelve specimens were identified from the locality HB 2030. Locality numbers are reported on the measured profiles (Figs. 3–7, 10). Measurements of diameter (in mm), followed by whorl height (H), whorl width (W), and umbilical diameter (U) are expressed as percentages of the shell diameter, and are plotted on bivariate diagrams for species with large samples.

Class Cephalopoda

Subclass Ammonoidea Zittel, 1884

Order Ceratitida Hyatt, 1884

Superfamily Sagecerataceae Hyatt, 1884

Family Sageceratidae Hyatt, 1884

Genus *Sageceras* Mojsisovics, 1873

***Sageceras walteri* Mojsisovics, 1882**

Pl. 16: 10; Pl. 18: 3–5

1882	*Sageceras walteri* – Mojsisovics, p. 187; Pl. 53: 9a–c [holotype].
? 1882	*Sageceras walteri* – Mojsisovics, p. 187; Pl. 53: 13.
non 1882	*Sageceras walteri* – Mojsisovics, p. 187; Pl. 53: 11, 12 [= *Sageceras haidingeri*].
1905	*Sageceras gabbi* – Hyatt & Smith, Pl. 74: 8, 9; Pl. 75: 14, 15.
1914	*Sageceras gabbi* – Smith, Pl. 11: 8, 9; Pl. 12: 14, 15; Pl. 21: 18–20.
non 1914	*Sageceras gabbi* – Smith, Pl. 6: 1–3 [= *Sageceras haidingeri*]
1982	*Sageceras walteri* – Silberling & Nichols, p. 18; Pl. 5: 10–12.
1989	*Sageceras* cf. *S. walteri* – Bucher, p. 962; Text-fig. 4; Pl. 4: 13, 14.

Description. – Involute shell, with a nearly closed umbilicus, whose proportion slightly increases in later whorls. Extremely compressed oxycone, with high whorl section, flattened, weakly bicarinate venter, and high, convex umbilical wall. Shell smooth or with ornamentation consisting only of sinuous growth lines. Suture line ceratitic and serial with numerous, narrow, elongated, U-shaped, and bifid lobes.

Measurements. – See Appendix.

Discussion. – This rare species occurs throughout most of the Anisian. Distinction from the Late Triassic *S. haidingeri* is quite tenuous (Silberling & Nichols 1982). However, our material confirms the distinction made by Spath (1934) in that *S. walteri* differs from *S. haidingeri* by having a narrower umbilicus at comparable diameters.

Occurrence. – Oliver Gulch (Augusta Mountains): HB 2041 (1); *Rotelliformis* Zone (Late Anisian). Rieber Gulch (Augusta Mountains): HB 2007 (1); *Weitschati* Zone (Late Anisian). Fossil Hill (Humboldt Range): FHB 8 (2); *Rotelliformis* Zone (Late Anisian).

Superfamily Ceratitaceae Mojsisovics, 1879

Family Acrochordiceratidae Arthaber, 1911

Genus *Acrochordiceras* Hyatt, 1877

***Acrochordiceras carolinae* Mojsisovics, 1882**

Pl. 2: 5–9

1882	*Acrochordiceras carolinae* – Mojsisovics, p. 141; Pl. 28: 14; Pl. 36: 3.
1982	*Acrochordiceras* cf. *A. carolinae* – Silberling & Nichols, p. 22; Pl. 5: 8, 9; Text-fig. 14.

Description. – Whorl section varies from circular to high oval. Broadly rounded venter, with indistinct ventral shoulders. Convex flanks, with greatest width low on flanks. High, convex umbilical wall. Ornamentation consisting of prorsiradiate and slightly sinuous, plicate ribs, developing greatest strength as they cross the venter. One to three ribs arise from tubercles at or just below mid-flank. Inner whorls involute, while outer whorls are moderately evolute with umbilical bullae replacing the tubercles of the inner whorls. Suture line subammonitic, with slightly denticulate saddles and deeply indented lobes.

Measurements. – See Appendix.

Discussion. – As noted by Silberling & Nichols (1982), species of *Acrochordiceras* from Nevada display a wide range of intraspecific variability, varying from compressed, non-tuberculate forms to depressed, bullate forms. Our specimens match the variability of *A. carolinae* and are referred to this species.

Occurrence. – Oliver Gulch (Augusta Mountains): HB 741 (3); *Shoshonensis* Zone (Middle Anisian). Ferguson Canyon (Augusta Mountains): HB 739 (5); *Shoshonensis* Zone (Middle Anisian). This species is rare in the *Mojsvari* Subzone where it has its latest occurrence, but it is very abundant in the older subzones of the *Shoshonensis* Zone.

Family Balatonitidae Spath, 1951

Genus *Platycuccoceras* Bucher, 1988

Platycuccoceras sp. indet.

Pl. 2: 3

? 1992b *Platycuccoceras* sp. indet. – Bucher, p. 432; Pl. 6: 22, 23.

Description. – Compressed whorl section, with flattened flanks and a low-arched venter. Ornamentation consisting of constrictions and weak ribs that cross the venter with a chevron shape but fade on mid-flank. No lateral, marginal, or ventral tuberculation.

Discussion. – A single fragment from the *Mojsvari* Subzone is referred to *Platycuccoceras* sp. indet. Absence of both lateral and ventral tubercles supports its assignment to *Platycuccoceras* rather than to *Balatonites*. It differs from the older *P. cainense* (*Ransomei* Subzone, *Shoshonensis* Zone) by its chevron-shaped ventral ribbing. It is similar to *Platycuccoceras* sp. indet. Bucher, 1992 (Pl. 6: 22, 23) from the *Ransomei* Subzone. Therefore, *Platycuccoceras* sp. indet. is the youngest occurrence of the genus in the Nevada sequence.

Occurrence. – Ferguson Canyon (Augusta Mountains): HB 739 (1); *Shoshonensis* Zone (Middle Anisian).

Genus *Balatonites* Mojsisovics, 1879

Balatonites hexatuberculatus n. sp.

Pl. 2: 1, 2?

Diagnosis. – *Balatonites* with six rows of tubercles and tuberculation predominant over ribbing.

Holotype. – PIMUZ 25134, Loc. HB 739, Ferguson Canyon (Augusta Mountains); *Mojsvari* Subzone, *Shoshonensis* Zone, Middle Anisian.

Etymology. – Species name refers to its six rows of tubercles.

Description. – Evolute coiling. Whorl section is fastigate, compressed, with flattened flanks and a tectiform venter. Six rows of tubercles: thin, bullate umbilical, lateral slightly below mid-flank, spinose lateral slightly above mid-flank, small but conspicuous inner ventrolateral, slightly spinose outer ventrolateral, and clavate ventral. Spinose lateral tubercles are outnumbered 2:1 by those of other rows. Tubercles are arranged on weak, thin, prorsiradiate, and slightly concave ribs. Suture line is unknown.

Likely inner whorls (~ 15 mm) are very evolute, with a fastigate, compressed whorl section and flattened flanks. Slightly sinuous, widely spaced, prorsiradiate ribs on innermost whorls, subsequently splitting into an umbilical tubercle and a rib on upper flank at a diameter of about 10 mm; some ribs bear weak lateral tubercles; ventral clavi are present at diameters greater than 7.5 mm, and are bordered by a row of marginal tubercles.

Measurements. – See Appendix.

Discussion. – Although the available material includes only one incomplete, distorted mature specimen and two early juvenile specimens, the six rows of tubercles visible on the adult shell distinguish this species from all other known congeneric species. This species differs from the older *B. shoshonensis* and *B. whitneyi* by its six rows of tubercles, and also from *B. shoshonensis* by the extension of the siphonal clavi to a larger size. Thus, *B. hexatuberculatus* may be descended from *B. whitneyi*. This species also differs from the European *Balatonites* (e.g. gr. *balatonicus*, gr. *egregius*) in that the lateral row of spinose tubercles is higher on the flanks, and by the greater number of rows of tubercles.

Occurrence. – Ferguson Canyon (Augusta Mountains): HB 739 (3); *Mojsvari* Subzone, *Shoshonensis* Zone (Middle Anisian).

Family Ceratitidae Mojsisovics, 1879
Subfamily Beyrichitinae Spath, 1934

Genus *Chiratites* n. gen.

Type species. – *Chiratites retrospinosus* n. sp.

Diagnosis. – Compressed, flat-sided beyrichitin with a low-arched venter. Inner whorls involute, smooth, or with sporadic weak sinuous folds. Transition to sub-mature and mature stages is marked by egressive coiling, thickening of the body chamber, and development of an inner ventrolateral row of widely spaced spines, and an occasional row of lateral tubercles.

Etymology. – Genus named after R. Chirat (Lyon).

Composition of the genus. – *Chiratites retrospinosus* n. sp. and *C. bituberculatus* n. sp.

Discussion. – *Chiratites* embodies a unique combination of characteristics quite unlike that of any previously known beyrichitin genus. The smooth, compressed, and involute inner whorls followed by the development of hollow marginal spines at later ontogenetic stages make

Fig. 14. Suture line (× 4) of *Chiratites retrospinosus* n. gen. n. sp. Paratype PIMUZ 25217, Loc. HB 739, Ferguson Canyon (Augusta Mountains); *Mojsvari* Subzone, *Shoshonensis* Zone, Middle Anisian.

the genus easily distinguishable. True spines, like those of *Chiratites*, were previously unknown among the Beyrichitinae.

Occurrence. – Ferguson Canyon (Augusta Mountains): *Mojsvari* Subzone, *Shoshonensis* Zone (Middle Anisian).

Chiratites retrospinosus n. sp.

Figs. 14–16; Pl. 3: 1–4

Diagnosis. – *Chiratites* with a single row of spines at maturity.

Holotype. – PIMUZ 25214, Loc. HB 739, Ferguson Canyon (Augusta Mountains); *Mojsvari* Subzone, *Shoshonensis* Zone, Middle Anisian.

Etymology. – Species name refers to its inner ventrolateral row of spines projected backward.

Description. – Large-sized beyrichitin with involute, compressed, and nearly smooth phragmocone, and spinose, egressive mature body chamber. Immature stages high whorled, involute, with a very narrow umbilicus, a low-arched venter, and nearly flat flanks. Ventral shoulder is

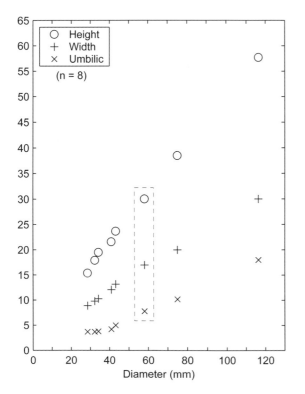

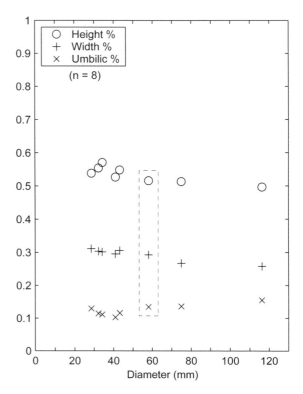

Fig. 15. Scatter diagram of H, W, and U, and of H/D, W/D, and U/D against corresponding diameter for the species of the new genus *Chiratites* (Augusta Mountains; *Mojsvari* Subzone, *Shoshonensis* Zone, Middle Anisian). All points represent *C. retrospinosus* n. gen. n. sp. except for those within the dashed frame, which correspond to *C. bituberculatus* n. gen. n. sp.

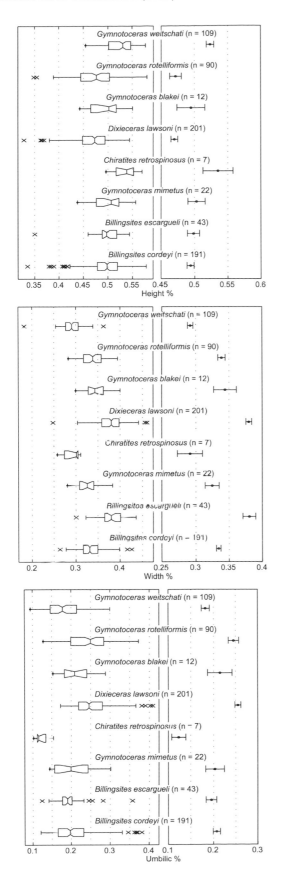

Fig. 16. Box and mean plots for the studied species of Beyrichitinae. See text for explanations.

narrow but rounded. Faint irregular, somewhat sinuous folds occur sporadically on flanks but do not cross the permanently smooth venter. Transition to submature and mature stages marked by egressive coiling starting at a diameter of about 45 mm, as well as by development of widely spaced, progressively longer, and backward-directed, hollow inner ventrolateral spines at a diameter of about 65 mm; these spines alternate on each side of the shell. Mature body chamber with high oval whorl section, higher arched venter, and subdued umbilical shoulders. Suture line has indented lobes and weakly indented saddles.

Measurements. – See Figures 15 and 16. Height (H) and width (W) of whorl section may have isometric growth, while the umbilical diameter (U) appears to have either allometric growth or an abrupt change in isometric growth at a diameter of about 43 mm. H and W tend to decrease proportionally as diameter (D) increases, while U becomes more or less stable at diameters greater than about 43 mm. It is noteworthy that *C. retrospinosus* is easily recognizable from all other studied beyrichitins by its significantly narrower umbilical diameter (Fig. 16), and by a less clear-cut compressed shell (low whorl width overlapping only with *Gymnotoceras weitschati*).

Discussion. – The type species differs from *C. bituberculatus* mainly by the absence of a lateral row of tubercles on middle whorls, and by the slightly greater size at which tuberculation develops.

Occurrence. – Ferguson Canyon (Augusta Mountains): HB 739 (10); *Mojsvari* Subzone, *Shoshonensis* Zone (Middle Anisian).

Chiratites bituberculatus n. sp.

Fig. 15; Pl. 2: 4

Diagnosis. – *Chiratites* with an inner ventrolateral row and a lateral row of tubercles.

Holotype. – PIMUZ 25213, Loc. HB 739, Ferguson Canyon (Augusta Mountains); *Mojsvari* Subzone, *Shoshonensis* Zone, Middle Anisian.

Etymology. – Species name refers to the two rows of tubercles.

Description. – Involute, high oval, compressed shell with a low-arched, relatively broad venter, and slightly convex flanks. Inner whorls are very involute and nearly smooth. Submature stage with slight egressive coiling, thin, weak ribs bearing two rows of tubercles (lateral and inner ventrolateral) beginning at a diameter of about 50 mm, and broadened venter. Suture line and body chamber are unknown.

Measurements. – The single specimen available for measurement falls within the range of *C. retrospinosus* (see Fig. 15).

Discussion. – *Chiratites bituberculatus* differs from *C. retrospinosus* by its more depressed whorl section, by the presence of lateral tubercles at mid-flank at the submature stage, by the earlier development of the inner ventrolateral spines, which do not tend to be directed backward, and by bearing weak ribs rather than sinuous folds as in *C. retrospinosus*. The smooth, very involute, and compressed inner whorls of both species are identical.

Occurrence. – Ferguson Canyon (Augusta Mountains): HB 739 (1); *Mojsvari* Subzone, *Shoshonensis* Zone (Middle Anisian).

Genus *Billingsites* n. gen.

Type species. – *Billingsites cordeyi* n. sp.

Diagnosis. – Involute, high-whorled, moderately compressed shell with a high oval, polygonal or subrectangular whorl section, a low-arched to subtabulate venter, convex flanks with greatest width just below mid-flank, narrow, rounded to slightly angular ventral shoulders, and steep umbilical wall. Ornamentation typically consisting of elongated, swollen lateral nodes (particularly well developed on middle whorls) just below mid-flank, of sinuous, thin to thick ribs slightly projected and fading on venter, and occasional faint marginal tubercles.

Etymology. – Genus named after W. Billings (Unionville).

Composition of the genus. – *Billingsites cordeyi* n. sp. and *B. escargueli* n. sp.

Discussion. – *Billingsites* differs from *Gymnotoceras* by its less involute coiling, by the absence of a rounded keel, by having a wider venter crossed by less projected growth lines, a more subrectangular whorl section, more widely spaced ribs, and the presence of lateral bullae. *Billingsites* differs from *Eogymnotoceras* by its more involute coiling, by the absence of a rounded keel, and by having lateral instead of umbilical bullae. *Billingsites* differs from *Anagymnotoceras* by its more involute coiling, its more subrectangular whorl section, and by having more projected growth lines on the venter. *Billingsites* differs from *Favreticeras* by its less egressive coiling at maturity, by the

presence of lateral bullae throughout most of ontogeny, and by having higher whorls. *Billingsites* differs from *Dixieceras* by its more involute coiling, by having a subrectangular whorl section, lateral instead of enlarged umbilical bullae, and higher whorls.

Species attributed to *Billingsites* have a wide range of intraspecific variability, which is in accordance with the First Buckman's Law of Covariation (Westermann 1966), i.e. a covariation of the stoutness of the ornamentation and the morphology of the whorl section (coiling and compression) throughout entire ontogeny. Therefore, comparison of *Billingsites* to Alpine typological taxa is difficult. Some variants of *Billingsites* are very close to the European *Schreyerites abichi* (Tatzreiter & Balini, 1993), but *Billingsites* always differs by having lateral bullae instead of small lateral tubercles, and by having a venter that is subtabulate rather than rounded on outer whorls.

Occurrence. – Oliver Gulch, Ferguson Canyon, Muller Canyon, Rieber Gulch (Augusta Mountains) and Fossil Hill (Humboldt Range); *Weitschati* and *Mimetus* zones (Late Anisian).

Billingsites cordeyi n. sp.

Figs. 11–13, 16–18; Pls. 5, 6; Pl. 7: 1–4, 6

Diagnosis. – *Billingsites* with dense ribbing, well-developed lateral bullae, and a relatively large adult size.

Holotype. – PIMUZ 25136, Loc. HB 590, Oliver Gulch (Augusta Mountains); *Cordeyi* Subzone, *Weitschati* Zone, Late Anisian.

Etymology. – Species named after F. Cordey (Lyon).

Description. – Involute coiling with slight egression on outer whorls. Compressed whorl section, varying from high oval on slender forms to subrectangular or subhexagonal on robust forms. Low-arched, rounded venter on slender forms to tabulate on robust forms. Flanks are convex, slightly converging towards marginal shoulders, which are slightly angular to conspicuous. Abrupt umbilical shoulder and low umbilical wall.

Ornamentation typically varies in strength within populations and consists of thin, slightly sinuous ribs to thick, straight ribs. Ribbing is slightly prorsiradiate, usually branching just below mid-flank. Slender variants have weak ribs rising just below mid-flank; these ribs gradually change to a bullate form as shell robustness increases; robust forms develop strong nodes at the branching point of ribs. Ribbing is projected forward on ventral shoulders; this projection leads to clavate-like forms on robust variants. The bullate/nodose lateral tuberculation begins at a diameter of about 20 mm.

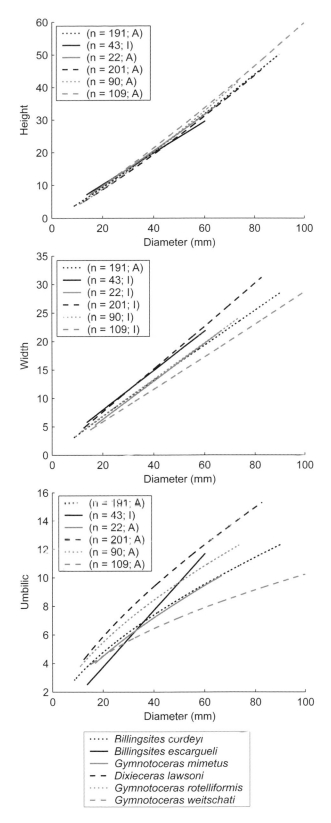

Fig. 17. Growth curves for the studied species of Beyrichitinae. I: isometric growth; A: allometric growth.

Suture line subammonitic, with four rounded or slightly indented saddles. Suture line becoming ammonitic at maturity.

Measurements. – See Figures 11, 16, and 17. H, W, and U have an expected normal distribution except for the near-adult whorls of the shell, which are characterized by egressive coiling, i.e. a tendency towards more evolute whorls (Fig. 11). The three parameters are characterized by allometric growth, which is nevertheless weak for whorl width (Fig. 17).

Discussion. – *Billingsites cordeyi* differs from *B. escargueli* by having denser, more sinuous ribs, a larger mature size, and a significantly more compressed shell (see Fig. 16).

Occurrence. – Oliver Gulch (Augusta Mountains): HB 583 (63), HB 584 (+300), HB 585 (154), HB 590 (32), HB 708 (10), HB 713 (1), HB 2052 (17); *Cordeyi* Subzone, *Weitschati* Zone (Late Anisian). Ferguson Canyon (Augusta Mountains): HB 740 (50), HB 750 (5), HB 751 (2?), HB 752 (1?), HB 754 (7), HB 2020 (1), HB 2021 (17), HB 2033 (22), HB 2034 (162), HB 2036 (13), HB 2053 (32); *Cordeyi* Subzone, *Weitschati* Zone (Late Anisian). Muller Canyon (Augusta Mountains): HB 735 (1?), HB 737 (43); *Cordeyi* Subzone, *Weitschati* Zone (Late Anisian). Rieber Gulch (Augusta Mountains): HB 2003 (98), HB 2004 (122), HB 2007 (1), HB 2012 (14), HB 2013 (16), HB 2014 (62), HB 2016 (4), HB 2017 (50), HB 2070 (2), HB 2072 (1); *Cordeyi* Subzone, *Weitschati* Zone (Late Anisian).

Billingsites escargueli n. sp.

Figs. 16, 17, 19, 20; Pl. 7: 5; Pl. 8; Pl. 9: 1–3

Diagnosis. – Thick-whorled *Billingsites* with strong lateral bullae, ribbing faint or absent, and smooth, compressed, very involute inner whorls.

Holotype. – PIMUZ 25166, Loc. HB 738, Ferguson Canyon (Augusta Mountains); *Cordeyi* Subzone, *Weitschati* Zone, Late Anisian.

Etymology. – Species named after G. Escarguel (Lyon).

Description. – Involute shell with a moderately compressed whorl section and a steep, low umbilical wall. Inner whorls with a high oval whorl section, a rounded venter, slightly convex flanks, and a nearly smooth phragmocone with occasional faint, weak folds. Outer whorls with a subrectangular whorl section, lateral bullae, slightly angular ventral shoulders, and a subtabulate venter. Thin, slightly prorsiradiate ribs branch from

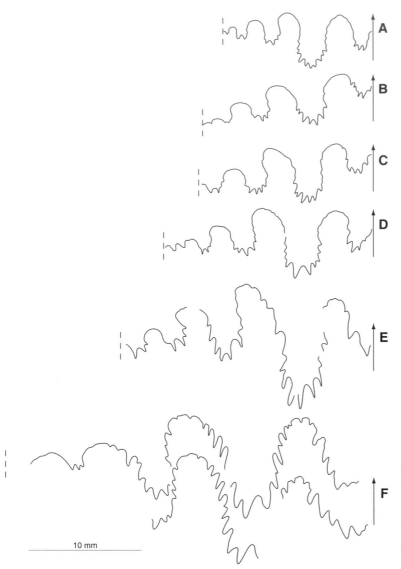

Fig. 18.   Suture lines (× 3) of *Billingsites cordeyi* n. gen. n. sp. A: Paratype PIMUZ 25143, Loc. HB 590, Oliver Gulch (Augusta Mountains); *Cordeyi* Subzone, *Weitschati* Zone, Late Anisian; reversed. B: Paratype PIMUZ 25161, Loc. HB 584, Oliver Gulch (Augusta Mountains); *Cordeyi* Subzone, *Weitschati* Zone, Late Anisian. C: Paratype PIMUZ 25163, Loc. HB 590, Oliver Gulch (Augusta Mountains); *Cordeyi* Subzone, *Weitschati* Zone, Late Anisian; reversed. D: Paratype PIMUZ 25162, Loc. HB 590, Oliver Gulch (Augusta Mountains); *Cordeyi* Subzone, *Weitschati* Zone, Late Anisian. E: PIMUZ 25159, Loc. HB 2033, Ferguson West (Augusta Mountains); *Cordeyi* Subzone, *Weitschati* Zone, Late Anisian; reversed. F: PIMUZ 25160, Loc. HB 740, Ferguson Canyon (Augusta Mountains); *Cordeyi* Subzone, *Weitschati* Zone, Late Anisian.

umbilical bullae, fade on upper flanks, and occasionally develop faint tubercles where projected forward on ventral shoulders. The bullate/nodose lateral tuberculation first begins at diameters varying from 10 mm to 25 mm. Suture line subammonitic with four weakly indented saddles becoming deeply crenulated at maturity.

Measurements. – See Figures 16, 17, and 20. As is the case for *Billingsites cordeyi*, the three measured parameters of *B. escargueli* have a normal distribution except for the near-adult whorls, which are characterized by egressive coiling (Fig. 20). This species appears to be unique because it displays highly significant isometric growth for the three parameters, in contrast to *B. cordeyi* whose parameters are allometric (Fig. 17). *B. escargueli* also differs from other studied beyrichitins by having a significantly thicker whorl width, which is only comparable to that of *Dixieceras lawsoni* (Fig. 16).

Discussion. – *Billingsites escargueli* differs from *B. cordeyi* by having a more subrectangular whorl section, weaker, less frequent ribs that fade on upper flanks, stronger lateral bullae, and nearly smooth innermost whorls.

Occurrence. – Ferguson Canyon (Augusta Mountains): HB 738 (82), HB 740 (2); *Cordeyi* Subzone, *Weitschati* Zone (Late Anisian). Rieber Gulch (Augusta Mountains): HB 2015 (2), HB 2016 (4); *Cordeyi* Subzone, *Weitschati* Zone (Late Anisian).

Genus *Dixieceras* n. gen.

Type species. – *Ceratites (Philippites) lawsoni* Smith, 1914

Diagnosis. – Moderately involute beyrichitin with strong, prorsiradiate umbilical bullae, a high, rounded, smooth venter, and a trapezoidal whorl section.

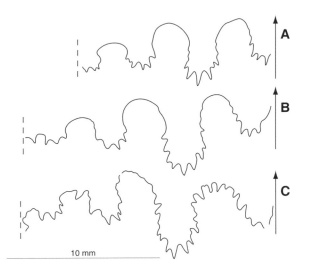

Fig. 19. Suture lines (× 4) of *Billingsites escargueli* n. gen. n. sp. A: Paratype PIMUZ 25174, Loc. HB 738, Ferguson Canyon (Augusta Mountains); *Cordeyi* Subzone, *Weitschati* Zone, Late Anisian; reversed. B: PIMUZ 25164, Loc. HB 2015, Rieber Gulch (Augusta Mountains); *Cordeyi* Subzone, *Weitschati* Zone, Late Anisian. C: Paratype PIMUZ 25180, Loc. HB 738, Ferguson Canyon (Augusta Mountains); *Cordeyi* Subzone, *Weitschati* Zone, Late Anisian.

Etymology. – Genus name derived from the Dixie Valley, Pershing County, Nevada.

Composition of the genus. – Type species only.

Description. – As for the type species.

Discussion. – *Dixieceras* differs from *Gymnotoceras* by its smooth, broadly rounded venter, by its absence of a keel, presence of enlarged and curved umbilical bullae, absence of megastriae, less projected growth lines on the venter, less involute coiling, and a more depressed, subtriangular whorl section. *Dixieceras* differs from *Eogymnotoceras* by absence of a keel, a subtriangular whorl section with rounded ventral shoulders, absence of parabolic nodes, and absence of megastriae. *Dixieceras* differs from *Anagymnotoceras*, to which it is superficially similar, by absence of parabolic nodes, presence of umbilical bullae also on outer whorls, a subtriangular whorl section also on outer whorls, higher whorls, and marked projection of growth lines on venter. *Dixieceras* differs from *Billingsites* by its coarser and more prorsiradiate bullae and a rounded venter. *Dixieceras* differs from *Favreticeras* by its subrectangular whorl section, the position of its umbilical bullae, and a more rounded venter. *Dixieceras* differs from *Frechites* by having a smooth venter without a keel throughout ontogeny, inflated, prorsiradiate umbilical bullae instead of lateral, narrow bullae, higher whorls, and a subtriangular whorl section. *Dixieceras* differs from *Parafrechites* by the absence of a keel, presence of enlarged

and curved umbilical bullae, and a more subtriangular whorl section.

Occurrence. – As for the type species.

Dixieceras lawsoni (Smith, 1914)

Figs. 16, 17, 21, 22; Pls. 10–12; Pl. 18: 1, 2

1914 *Ceratites (Philippites) lawsoni* – Smith, p. 108; Pl. 56: 1–5 [holotype], 6–13; Pl. 57: 1–17.
1934 *Frechites* sp. – Spath, p. 418.
1982 ? *Frechites occidentalis* – Silberling & Nichols, p. 30.

Description. – Moderately involute, high-whorled shell with a slightly overhanging umbilical wall, a well-rounded, low-arched, broad, smooth venter, and slightly angular ventral shoulders. Whorl section varies from high oval, compressed, with gently converging flanks, to subtriangular and depressed. On inner whorls, ornamentation varies from nearly smooth, with weak ribs, to dense, prorsiradiate ribs projected on the venter. On outer whorls, ornamentation varies from weak, sinuous ribs, with intercalated or branching ribs, to strong, prorsiradiate umbilical bullae from which arise two strong ribs projected forward at the ventral shoulder. Umbilical bullae appear at a diameter of about 35 mm, and then strengthen throughout growth, thus creating a subtriangular whorl section. Suture line ceratitic, with three well-rounded, smooth saddles.

Measurements. – See Figures 16, 17, and 22, as well as the Appendix. Except for the inner whorls, H, W, and U have a normal distribution. The three parameters also appear to be quite variable with allometric growth for whorl height and umbilical diameter, and isometric growth for whorl width. This greater variability is reflected by forms ranging from weakly bullate and compressed to highly bullate and depressed (e.g. Pl. 10: 4, 7). Samples from Fossil Hill and the Augusta Mountains are statistically similar in mean and variance.

Discussion. – Silberling & Nichols (1982) followed the proposal of Spath (1934) in referring "*Ceratites*" *lawsoni* Smith, 1914 to *Frechites* and tentatively to *F. occidentalis*, although they did not document this species in stratigraphically controlled collections. Our new collections from both the Augusta Mountains and the Fossil Hill type area reveal that "*Ceratites*" *lawsoni* is a valid morphological entity and that its stratigraphical occurrence is much earlier than that of *Frechites*. Furthermore, our bed-by-bed sampled assemblages of "*Ceratites*" *lawsoni* also reveal that its curved and inflated umbilical bullae enable clear distinction from *Frechites* and is thus deserving of

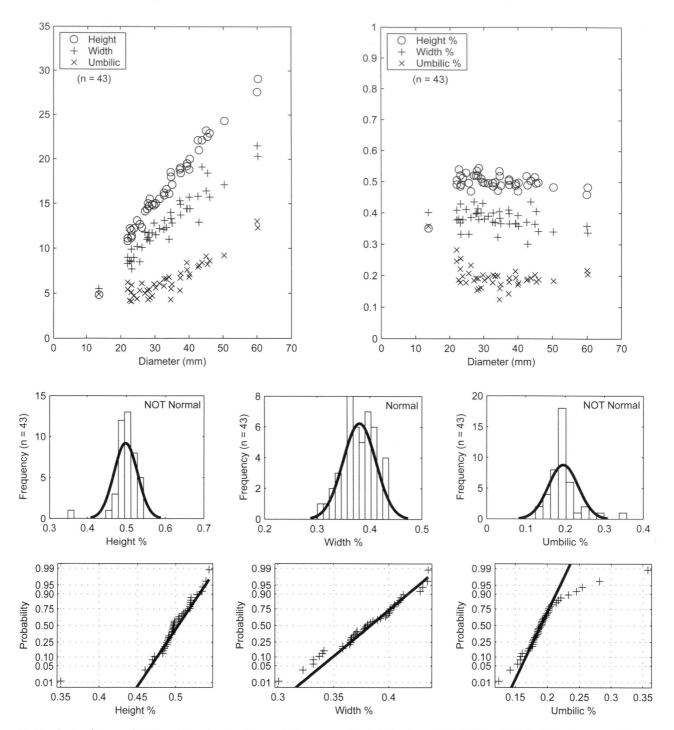

Fig. 20.   Scatter diagram of H, W, and U, and scatter diagram, histograms, and probability plots of H/D, W/D, and U/D for *Billingsites escargueli* n. gen. n. sp. (Augusta Mountains; *Cordeyi* Subzone, *Weitschati* Zone, Late Anisian).

separate status. Since this species displays such distinctive characteristics when compared to all known Anisian beyrichitins, its erection to a new genus is fully justified.

Occurrence. – Oliver Gulch (Augusta Mountains): HB 597 (53), HB 710 (7), HB 711 (4), HB 2040 (3); *Lawsoni* Subzone, *Mimetus* Zone (Late Anisian). Ferguson Canyon (Augusta Mountains): HB 2026 (1), HB 2027 (2), HB 2032 (11), HB 2037 (14); *Lawsoni* Subzone, *Mimetus* Zone (Late Anisian). Muller Canyon (Augusta Mountains): HB 717 (6), HB 735 (24); *Lawsoni* Subzone, *Mimetus* Zone (Late Anisian). Fossil Hill (Humboldt Range): FHB 8 (100), FHB 9 (35); *Lawsoni* Subzone, *Mimetus* Zone (Late Anisian).

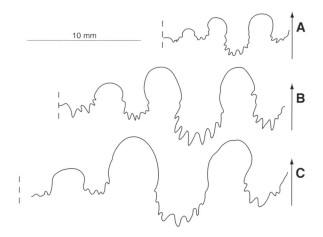

Fig. 21. Suture lines (× 3, reversed) of *Dixieceras lawsoni* (Smith, 1914). A: PIMUZ 25255, Loc. FHB 9, Fossil Hill (Humboldt Range); *Lawsoni* Subzone, *Mimetus* Zone, Late Anisian. B: PIMUZ 25256, Loc. HB 710, Oliver Gulch (Augusta Mountains); *Lawsoni* Subzone, *Mimetus* Zone, Late Anisian. C: PIMUZ 25254, Loc. HB 717, Muller Canyon (Augusta Mountains); *Lawsoni* Subzone, *Mimetus* Zone, Late Anisian.

Genus *Gymnotoceras* Hyatt, 1877

Gymnotoceras rotelliformis Meek, 1877

Figs. 16, 17, 23, 24; Pl. 13

1877   *Gymnotoceras rotelliforme* – Meek, p. 111; Pl. 10: 9, 9a.

non 1904 *Beyrichites rotelliformis* – Smith, p. 379; Pl. 43: 13, 14; Pl. 45: 5 [= *Gymnotoceras blakei*]

1905   *Beyrichites rotelliformis* – Hyatt & Smith, p. 155; Pl. 23: 1–7a; Pl. 58: 5, 6.

? 1905   *Beyrichites rotelliformis* – Hyatt & Smith, p. 155; Pl. 58: 1–4, 7–15.

1914   *Beyrichites rotelliformis* – Smith, p. 118 [part]; Pl. 4: 1–7a; Pl. 8: 5–6; Pl. 14: 9–9a; Pl. 31: 3–4; Pl. 91: 1–2, 5–7.

? 1914   *Beyrichites rotelliformis* – Smith, p. 118; Pl. 8: 1–4, 7–15; Pl. 31: 1–2, 5–6; Pl. 91: 3–4, 8–10.

non 1914 *Beyrichites rotelliformis* – Smith, p. 118; Pl. 31: 1–2 [= *Gymnotoceras mimetus*]

1914   *Ceratites (Philippites) argentarius* – Smith, p. 107; Pl. 63: 1–3 [holotype], 4–11.

? 1914   *Ceratites (Philippites) argentarius* – Smith, p. 107; Pl. 63: 12–14.

1914   *Beyrichites tenuis* – Smith, p. 119; Pl. 32: 1–2 [holotype], 3–4; Pl. 89: 15–20.

? 1914   *Beyrichites tenuis* – Smith, p. 119; Pl. 32: 5–6.

1969   *Gymnotoceras deleeni* – McLearn, p. 24 [part]; Pl. 4: 1, 6; Pl. 5: 5.

1982   *Gymnotoceras rotelliformis* – Silberling & Nichols, p. 26; Pl. 7: 1–27; Pl. 8: 1–5 [revised].

1994   *Gymnotoceras smithi* – Tozer, p. 118; Pl. 59: 12, 13, 44a.

Description. – Very involute, compressed shell with a high oval whorl section, rounded ventral shoulders, a low-arched and narrow venter, which on robust variants bears a low rounded keel fading in late ontogenetic stages, convex flanks converging towards the venter, and a shallow umbilical wall. Dense, sinuous, prorsiradiate ribs highly projected on the venter and commonly branching from swollen primary ribs just below mid-flank. Primary ribs are outlined by megastriae on the phragmocone. Suture line subammonitic with four rounded and finely crenulated saddles. Suture line becoming ammonitic at maturity.

Measurements. – See Figures 16, 17, and 24. H, W, and U have a normal distribution throughout ontogeny. Whorl width is characterized by isometric growth, whereas whorl height and umbilical diameter display allometric growth. The two species *G. rotelliformis* and *G. weitschati* have rather similar growth curves except for the umbilical diameter, which is always larger in *G. rotelliformis*; in other words, at a similar diameter, *G. rotelliformis* differs from *G. weitschati* by its larger umbilical diameter. Note that the two species also have a significantly different mean for each parameter, but their distributions partially overlap.

Discussion. – This species has been revised by Silberling & Nichols (1982). Our material matches the intraspecific variation illustrated by these authors. It also highlights the previously unknown increasing frilling of the suture line at maturity. The slightly younger species *G. blakei* differs from *G. rotelliformis* by the presence of ribs that persist on the upper flanks, by higher rib density, by fewer branching ribs, and by a narrower venter at maturity. The older species *G. weitschati* differs by its greater adult size and a subtriangular whorl section with flattened flanks in intermediate stages. *G. mimetus* differs from *G. rotelliformis* by its weaker keel, a flattened whorl section on outer whorls, a broader venter, and weaker ornamentation.

Tozer (1994) assigned the specimens of *Gymnotoceras* of the *Deleeni* Zone from British Columbia to a new species (*G. smithi*) by assuming that *G. rotelliformis* has a more inflated and a "trifle blunter" venter. In fact, *G. smithi* is morphologically similar to *G. rotelliformis*, and its geometrical parameters (H, W, U) fall completely within the range of our samples of *G. rotelliformis*. We therefore consider *G. smithi* as a junior synonym of *G. rotelliformis*.

Occurrence. – Oliver Gulch (Augusta Mountains): HB 703 (2), HB 712 (6), HB 742 (17), HB 745 (12), HB 746 (3),

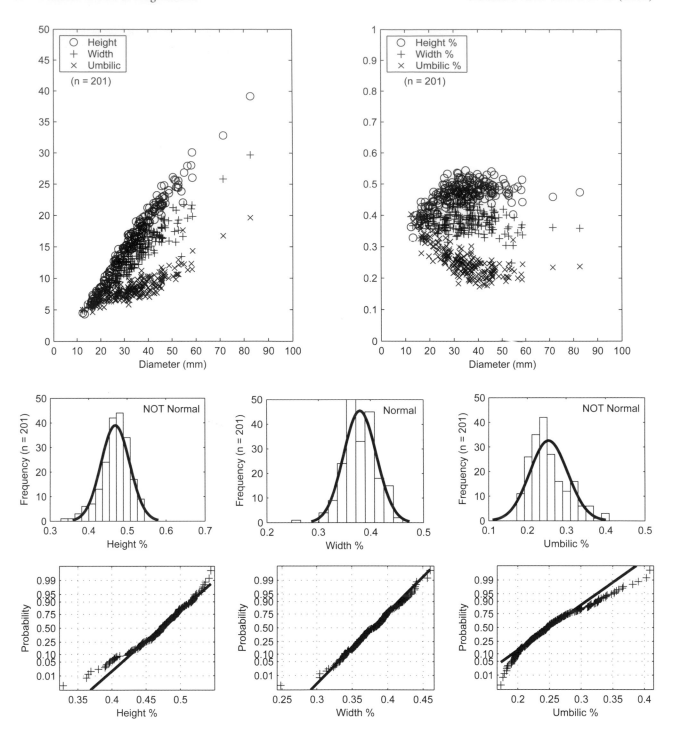

Fig. 22. Scatter diagram of H, W, and U, and scatter diagram, histograms, and probability plots of H/D, W/D, and U/D for *Dixieceras lawsoni* (Smith, 1914) (Augusta Mountains; *Lawsoni* Subzone, *Mimetus* Zone, Late Anisian).

HB 747 (3), HB 748 (1), HB 749 (1), HB 2041 (1), HB 2042 (1), HB 2044 (1); *Rotelliformis* Zone (Late Anisian). Muller Canyon (Augusta Mountains): HB 736 (2); *Rotelliformis* Zone (Late Anisian). Ferguson Canyon (Augusta Mountains): HB 2023 (7); *Rotelliformis* Zone (Late Anisian). Favret Canyon (Augusta Mountains): HR

block B (3); *Rotelliformis* Zone (Late Anisian). McCoy Mine (New Pass Range): HB 2063 (10); *Rotelliformis* Zone (Late Anisian). Fossil Hill (Humboldt Range): FHB 12 (4), FHB 13 (7), FHB 14 (10), FHB 15 (6), FHB 15A (15), FHB 16 (11), FHB 17 (15), FHB 18 (42), FHB 20 (3), FHB 22 (38), FHB 23 (2); *Rotelliformis* Zone (Late Anisian).

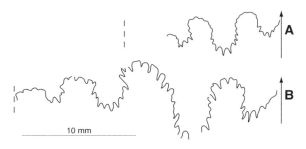

Fig. 23. Suture lines (× 3) of *Gymnotoceras rotelliformis* Meek, 1877. A: PIMUZ 25291, Loc. FHB 22, Fossil Hill (Humboldt Range); *Vogdesi* Subzone, *Rotelliformis* Zone, Late Anisian; reversed. B: PIMUZ 25281, Loc. FHB 18, Fossil Hill (Humboldt Range); *Vogdesi* Subzone, *Rotelliformis* Zone, Late Anisian.

Gymnotoceras blakei (Gabb, 1864)

Figs. 16, 25; Pl. 18: 6–7

1864	*Ammonites blakei* – Gabb, p. 24; Pl. 4: 14–15.
non 1877	*Gymnotoceras blakei* – Meek, p. 113; Pl. 10: 10–10a, 10b [= *Parafrechites meeki*]; Pl. 11: 6–6a [= *Frechites nevadanus*].
non 1904	*Ceratites (Gymnotoceras) blakei* – Smith, p. 386; Pl. 43: 9–10; Pl. 44: 2–3 [= *Parafrechites meeki*].
1905	*Ceratites (Gymnotoceras) blakei* – Hyatt & Smith, p. 173; Pl. 22: 10–11.
? 1905	*Ceratites (Gymnotoceras) blakei* – Hyatt & Smith, p. 173; Pl. 22: 12–23.
non 1905	*Ceratites (Gymnotoceras) blakei* – Hyatt & Smith, p. 173; Pl. 22: 1–5, 7–9 [= *Parafrechites meeki*].
1914	*Ceratites (Gymnotoceras) blakei* – Smith, p. 109; Pl. 3: 10–11; Pl. 16: 8–10, 17–19; Pl. 66: 1–2.
? 1914	*Ceratites (Gymnotoceras) blakei* – Smith, p. 109; Pl. 3: 12–23; Pl. 66: 3–8.
non 1914	*Ceratites (Gymnotoceras) blakei* – Smith, p. 109; Pl. 14: 10b; Pl. 65: 14–16 [= *Parafrechites meeki*].
1914	*Beyrichites falciformis* – Smith, p. 116; Pl. 91: 11–13; Pl. 92: 1–8.
1982	*Gymnotoceras blakei* – Silberling & Nichols, p. 26; Pl. 8: 6–19; Pl. 9: 1–18 [revised].

Description. – Involute, compressed, high-whorled beyrichitin with high oval whorl section, rounded ventral shoulders, convex flanks converging towards the venter, a low umbilical wall, and a narrowly rounded venter, which on robust variants bears a low, rounded keel fading in later ontogenetic stages. Ornamentation consisting of numerous, sinuous, prorsiradiate ribs, occasionally branching on lower flanks. Ribbing is of uniform strength from umbilicus to venter. Primary ribs outlined by megastriae. Ribs projected on the venter.

Measurements. – See Figures 16 and 25. Few specimens are available, but the measurements of H, W, and U are similar to those of *G. rotelliformis*.

Discussion. – The species has been revised by Silberling & Nichols (1982). It differs from *G. rotelliformis* by having denser, regular sinuous ribbing, and a narrower venter. The older species *G. weitschati* differs by its larger adult size and a subtriangular whorl section with flat, parallel flanks on outer whorls. The morphology of *G. blakei* is unique in that its robust variants are similar to *G. rotelliformis*, while its slender variants are similar to *G. weitschati*. *G. blakei* differs from *G. mimetus* by its convex flanks, a narrower venter, and coarser, denser ribs.

Occurrence. – Fossil Hill (Humboldt Range): FHB 28 (?), FHB 30 (8), FHB 31 (3), FHB 32 (8), FHB 34 (6); *Rotelliformis* Zone (Late Anisian).

Gymnotoceras weitschati n. sp.

Figs. 16, 17, 26, 27; Pls. 14, 15; Pl. 16: 1–9

Diagnosis. – Involute *Gymnotoceras* with a very narrow umbilicus, a subtriangular whorl section, a narrow venter, and sinuous ribs strengthening on upper flanks.

Holotype. – PIMUZ 25292, Loc. HB 2006, Rieber Gulch (Augusta Mountains); *Cordeyi* Subzone, *Weitschati* Zone, Late Anisian.

Etymology. – Species named after W. Weitschat (Hamburg).

Description. – Involute shell, with a very narrow umbilicus, and a high oval, compressed whorl section. Convex flanks converging towards the narrow and smooth venter, which occasionally bears a weak, low-rounded keel on inner whorls of robust variants. Inner whorls with dense, thin, sinuous, slightly prorsiradiate ribs of equal strength on the flanks. Outer whorls with flexuous ribs fading on lower flanks. Ribs occasionally branching near the umbilicus. Latest whorls smooth. Suture line subammonitic with four rounded and finely indented saddles. Suture line becoming ammonitic at maturity, with crenulated saddles.

Measurements. – See Figures 16, 17, and 27. H, W, and U have a normal distribution except for the whorl height, which departs from normality. This departure does not originate from differences between the various localities, subzones, or sections. Whorl height and umbilical width have allometric growth, whereas whorl width conforms

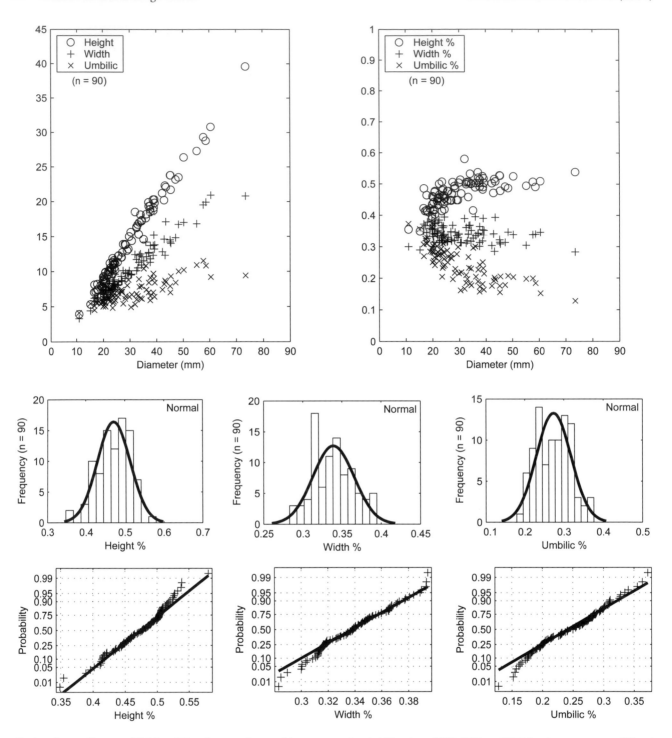

Fig. 24.   Scatter diagram of H, W, and U, and scatter diagram, histograms, and probability plots of H/D, W/D, and U/D for *Gymnotoceras rotelliformis* Meek, 1877 (Augusta Mountains; *Rotelliformis* Zone, Late Anisian).

to isometric growth. This species has a significantly different mean for each parameter in comparison to *G. rotelliformis*, even if their distributions partially overlap. Although this species is known from several successive layers, the three measures reveal no significant changes through time and space, and they have a narrow variability in comparison with *G. rotelliformis* or *D. lawsoni*.

Discussion. – This species differs from the overlying *G. rotelliformis* and *G. blakei* by having a more triangular whorl section, a narrower venter, a sinuous ribbing that fades on lower flanks, a keel visible only on inner whorls of very robust variants, and a larger adult size. The much older *G. ginsburgi* (*Shoshonensis* Zone) differs by having more compressed inner whorls with a subrectangular

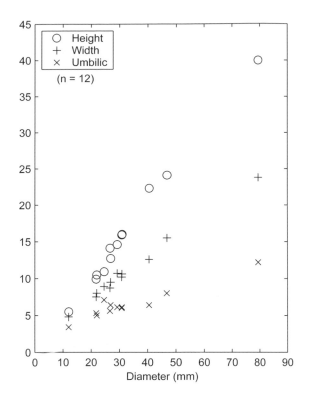

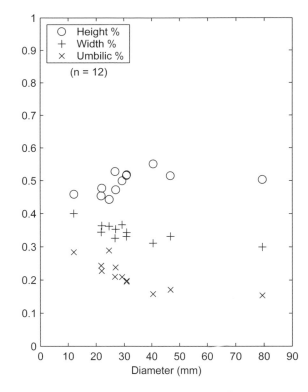

Fig. 25. Scatter diagram of H, W, and U, and of H/D, W/D, and U/D for *Gymnotoceras blakei* (Gabb, 1864) (Augusta Mountains; *Blakei* Subzone, *Rotelliformis* Zone, Late Anisian).

whorl section. *G. praecursor* (*Shoshonensis* Zone) differs by having a discoidal, compressed, and weakly ribbed phragmocone. *G. weitschati* differs from *G. mimetus* by having a subtriangular whorl section on outer whorls, ribs strengthening on upper flanks, and a larger adult size.

Occurrence. – Rieber Gulch (Augusta Mountains): HB 2003 (5), HB 2004 (13), HB 2006 (33), HB 2007 (11), HB 2009 (10), HB 2010 (49), HB 2013 (1), HB 2014 (2), HB 2050 (28), HB 2071 (1), HB 2072 (1); *Weitschati* Zone (Late Anisian). Ferguson Canyon (Augusta Mountains): HB 738 (1), HB 750 (3), HB 751 (3), HB 2021 (5?), HB 2030 (13), HB 2034 (5), HB 2053 (1); *Weitschati* Zone (Late Anisian). Oliver Gulch (Augusta Mountains): HB 584 (2), HB 585 (4), HB 596 (28), HB 597 (6), HB 706 (3), HB 708 (2), HB 713 (3), HB 749 (3?), HB 2052 (1); *Weitschati* Zone (Late Anisian). Muller Canyon (Augusta Mountains): HB 730 (5), HB 734 (15), HB 735 (9), HB 737 (1); *Weitschati* Zone (Late Anisian).

Gymnotoceras mimetus n. sp.

Figs. 16, 17, 28, 29; Pl. 17: 1–9

1914 *Beyrichites rotelliformis* – Smith, p. 118; Pl. 31: 1–2.

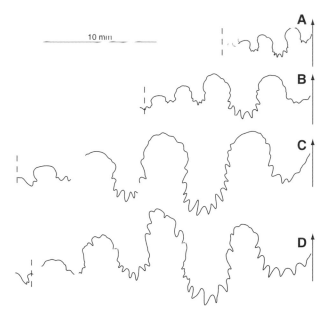

Fig. 26. Suture lines (× 3, reversed) of *Gymnotoceras weitschati* n. sp. A: Paratype PIMUZ 25299, Loc. HB 2006, Rieber Gulch (Augusta Mountains); *Cordeyi* Subzone, *Weitschati* Zone, Late Anisian. B: Paratype PIMUZ 25293, Loc. HB 2006, Rieber Gulch (Augusta Mountains); *Cordeyi* Subzone, *Weitschati* Zone, Late Anisian. C: Holotype PIMUZ 25292, Loc. HB 2006, Rieber Gulch (Augusta Mountains); *Cordeyi* Subzone, *Weitschati* Zone, Late Anisian. D: PIMUZ 25320, Loc. HB 2009, Rieber Gulch (Augusta Mountains); *Cordeyi* Subzone, *Weitschati* Zone, Late Anisian.

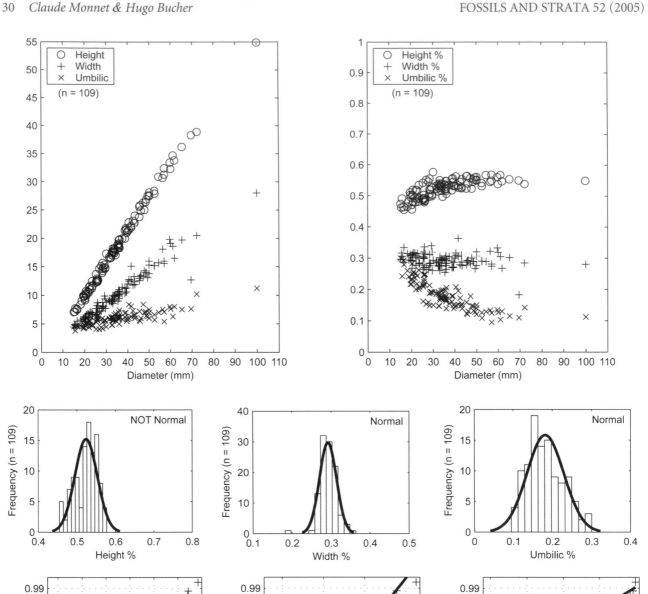

Fig. 27.   Scatter diagram of H, W, and U, and scatter diagram, histograms, and probability plots of H/D, W/D, and U/D for *Gymnotoceras weitschati* n. sp. (Augusta Mountains; *Weitschati* Zone, Late Anisian).

Fig. 28.   Suture line (× 4) of *Gymnotoceras mimetus* n. sp. Paratype PIMUZ 25267, Loc. FHB 9, Fossil Hill (Humboldt Range); *Lawsoni* Subzone, *Mimetus* Zone, Late Anisian.

Diagnosis. – *Gymnotoceras* with a small adult size, a compressed, flat-sided whorl section, and weak ornamentation with flexuous, relatively distant, thin ribs.

Holotype. – PIMUZ 25267, Loc. FHB 9, Fossil Hill (Humboldt Range); *Lawsoni* Subzone, *Mimetus* Zone, Late Anisian.

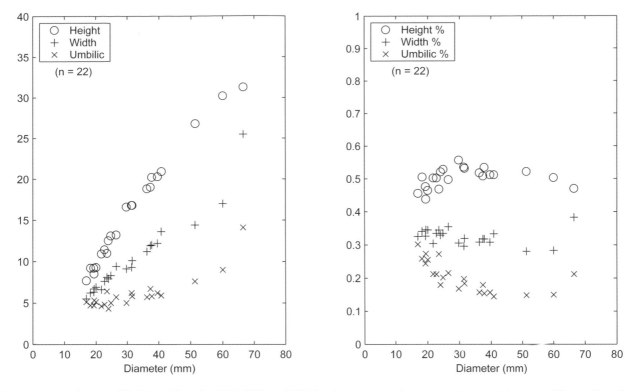

Fig. 29. Scatter diagram of H, W, and U, and of H/D, W/D, and U/D for *Gymnotoceras mimetus* n. sp. (Augusta Mountains; *Mimetus* Zone, Late Anisian).

Etymology. – Species name refers to its outer whorls, which are similar to those of *Billingsites* and *Nicholsites*.

Description. – Very involute coiling with slight egression of outer whorls. Whorl section high oval and compressed with a rounded venter; ventral shoulders rounded; flanks slightly convex on inner whorls, flat and subparallel on outer whorls; umbilical shoulders abrupt. Ornamentation consisting of fine, sinuous to flexuous, slightly prorsiradiate ribs, swelling on mid-flank of outer whorls into a bullate shape. Ribs projected on ventral shoulders. Suture line subammonitic with four rounded, finely crenulated saddles.

Measurements. – See Figures 16, 17, and 29. Whorl height and umbilical diameter display allometric growth, while whorl width is characterized by isometric growth.

Discussion. – *G. mimetus* differs from *G. rotelliformis* by having a weaker keel, a flattened whorl section on outer whorls, a broader venter, and weaker ornamentation. *G. blakei* differs from *G. mimetus* by having convex flanks, a narrower venter, and coarser, denser ribs. *G. weitschati* differs from *G. mimetus* by having a subtriangular whorl section on outer whorls, strengthened ribs on upper flanks, and a larger adult size.

Submature whorls of *G. mimetus* resemble those of *Billingsites* by having a compressed and flattened whorl section, and faint lateral bullae. However, *G. mimetus* differs from *Billingsites* by having keeled inner whorls, more projected growth lines on the venter, and more flexuous ribbing.

G. mimetus is strikingly similar to *Nicholsites parisi* Bucher, 1994 from the *Taylori* Zone (Middle Anisian), but its low, rounded keel on the inner whorls of robust variants can distinguish it from the latter.

Occurrence. – Fossil Hill (Humboldt Range): FHB 6 (4), FHB 8 (4), FHB 9 (17), FHB 11 (15); *Lawsoni* and *Spinifer* subzones, *Mimetus* Zone (Late Anisian). Oliver Gulch (Augusta Mountains): HB 597 (1); *Lawsoni* Subzone, *Mimetus* Zone (Late Anisian). Muller Canyon (Augusta Mountains): HB 735 (9); *Lawsoni* Subzone, *Mimetus* Zone (Late Anisian).

Subfamily Paraceratitinae Silberling, 1962

Genus *Bulogites* Arthaber, 1912

Bulogites mojsvari (Arthaber, 1896)

Fig. 30; Pl. 4: 1–4

1896 *Ceratites mojsvari* – Arthaber, p. 50; Pl. 4: 6.

Fig. 30. Suture line (×4) of *Bulogites mojsvari* (Arthaber, 1896). PIMUZ 25204, Loc. HR ##A, McCoy Mine (New Pass Range); *Mojsvari* Subzone, *Shoshonensis* Zone, Middle Anisian.

1992b *Bulogites* cf. *B. mojsvari* – Bucher, p. 437; Text-figs. 17–18.

Description. – High-whorled, moderately involute shell with a subrectangular whorl section, slightly convex, converging flanks, a low-arched venter, slightly angular ventral shoulders, a steep umbilical wall, and a narrow, angular umbilical shoulder. Ornamentation is trituberculate (blunt umbilical, nodose lateral just below mid-flank, and clavate marginal). Ribs are sinuous and slightly prorsiradiate. Intercalatory ribs may occur between branching ribs, which arise from lateral nodes. Ribs are coarser just above mid-flank, but fade on lower flanks. Dense and regularly spaced ribs crossing the venter in a U shape. Ribs become irregularly spaced and coarser on the body chamber. Suture line is ceratitic with crudely indented lobes.

Measurements. – See Appendix.

Discussion. – New material confirms the presence of *Bulogites* in Nevada, which was previously known only from a single float specimen (Bucher 1992b; re-illustrated here: Pl. 4: 2). The newly documented specimens pinpoint the stratigraphical position of the species in the Nevada sequence. It occurs in a distinct horizon belonging to the *Shoshonensis* Zone in the Augusta Mountains and in the McCoy Mine area (Fig. 1), as indicated by associated genera (*Acrochordiceras, Platycuccoceras* and *Balatonites*).

Occurrence. – Ferguson Canyon (Augusta Mountains): HB 739 (3); *Mojsvari* Subzone, *Shoshonensis* Zone (Middle Anisian). Muller Canyon (Augusta Mountains): USNM 452800 (1); *Mojsvari* Subzone, *Shoshonensis* Zone (Middle Anisian). McCoy Mine (New Pass Range): HR ##A (1); *Mojsvari* Subzone, *Shoshonensis* Zone (Middle Anisian).

Genus *Jenksites* n. gen.

Type species. – *Jenksites flexicostatus* n. sp.

Diagnosis. – Involute shell, with egressive coiling at maturity, a subrectangular whorl section, dense, flexuous ribs projected on the angular ventral shoulders, where they may bear small, faint tubercles, and without lateral tubercles.

Etymology. – Genus named after J. Jenks (Salt Lake City).

Composition of the genus. – Type species only.

Description. – As for the type species.

Discussion. – *Jenksites* differs from the co-occurring *Rieppelites* mainly by the absence of lateral tubercles and by having weaker marginal tubercles. *Jenksites* differs from *Bulogites* by having more flexuous ribs and by the absence of umbilical and lateral tubercles. *Jenksites* differs from *Rieberites* by the absence of lateral tubercles, the absence of a keel, by having flexuous rather than biconcave ribs, and a low-arched rather than subtabulate venter.

Occurrence. – As for the type species.

Jenksites flexicostatus n. sp.

Figs. 31–34; Pl. 19

Holotype. – PIMUZ 25321, Loc. HB 598, Oliver Gulch (Augusta Mountains); *Cordeyi* Subzone, *Weitschati* Zone, Late Anisian.

Etymology. – Species name refers to its flexuous ribbing.

Description. – Involute, compressed species with a subrectangular whorl section, a low-arched venter, angular ventral and umbilical shoulders, and a low umbilical wall. Coiling is egressive at maturity. Ornamentation consists of dense and regularly spaced, prorsiradiate, flexuous ribs. On inner whorls, thick primary ribs branch off on lower flanks. On outer whorls, the branching point shifts to a position just below mid-flank. Ribs projected on ventral shoulders, where they strengthen and may bear small, faint tubercles. On outer whorls, ribs fade or disappear

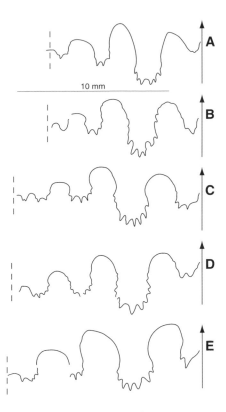

Fig. 31. Suture lines (×4) of *Jenksites flexicostatus* n. gen. n. sp. A: PIMUZ 25325, Loc. HB 713, Oliver Gulch (Augusta Mountains); *Cordeyi* Subzone, *Weitschati* Zone, Late Anisian; reversed. B: PIMUZ 25327, Loc. HB 2007, Rieber Gulch (Augusta Mountains); *Cordeyi* Subzone, *Weitschati* Zone, Late Anisian. C: PIMUZ 25326, Loc. HB 713, Oliver Gulch (Augusta Mountains); *Cordeyi* Subzone, *Weitschati* Zone, Late Anisian; reversed. D: PIMUZ 25329, Loc. HB 2007, Rieber Gulch (Augusta Mountains); *Cordeyi* Subzone, *Weitschati* Zone, Late Anisian. E: PIMUZ 25336, Loc. HB 713, Oliver Gulch (Augusta Mountains); *Cordeyi* Subzone, *Weitschati* Zone, Late Anisian.

completely on lower flanks. Suture line is ceratitic with narrow saddles and slightly crenulated flanks of saddles.

Measurements. – See Figures 32, 33, and 34. H, W, and U are highly variable and display allometric growth.

Discussion. – This species displays intraspecific variability ranging from variants with dense, thin, single ribs to variants with thick ribs branching below mid-flank and with faint tubercles on the ventral shoulders.

Occurrence. – Oliver Gulch (Augusta Mountains): HB 592 (6), HB 593 (1), HB 598 (4), HB 713 (8); *Cordeyi* Subzone, *Weitschati* Zone (Late Anisian). Ferguson Canyon (Augusta Mountains): HB 2020 (2), HB 2022 (1?); *Cordeyi* Subzone, *Weitschati* Zone (Late Anisian). Rieber Gulch (Augusta Mountains): HB 2007 (58); *Cordeyi* Subzone, *Weitschati* Zone (Late Anisian). McCoy Mine (New Pass Range): HB 2062 (7); *Cordeyi* Subzone, *Weitschati* Zone (Late Anisian).

Genus *Rieppelites* n. gen.

Type species. – *Rieppelites boletzkyi* n. sp.

Diagnosis. – Involute, compressed, high-whorled shell with a subrectangular to subhexagonal whorl section, a smooth, low-arched venter, angular lateral and umbilical shoulders, two rows of tubercles (lateral and marginal), and flexuous, rectiradiate ribs enhanced on ventral shoulders. Innermost whorls bear lateral parabolic nodes.

Etymology. – Genus named after O. Rieppel (Chicago).

Composition of the genus. – *Rieppelites boletzkyi* n. sp. and *R. shevyrevi* n. sp.

Discussion. – *Rieppelites* is closely related to *Jenksites*, from which it differs by having lateral tuberculation and more pronounced marginal tubercles. *Rieppelites* differs from the younger *Rieberites* by having flexuous rather than biconcave ribs, by having a low-arched venter, and absence of a keel.

Intraspecific variation of *Rieppelites* is characterized by a variation in the robustness of the lateral tubercles, which may completely disappear. Therefore, it may be difficult to differentiate between slender variants of *Rieppelites* and *Jenksites*, but robust variants of *Jenksites* have thicker, dichotomous ribs, which do not develop lateral tubercles as compared to robust variants of *Rieppelites*.

Rieppelites also exhibits certain affinities with several Alpine genera such as *Lardaroceras* Balini, 1992a, *Ronconites* Balini, 1992b, and *Pisaites* Balini, 1992b. *Rieppelites* differs from *Lardaroceras* and *Pisaites* by having parabolic nodes on the inner whorls and a smooth venter without a low, rounded keel. *Rieppelites boletzkyi* closely resembles *Ronconites tridentinus* Balini, 1992b. However, the two genera diverge in their late ontogenetic stages. *Ronconites* has a subquadrate whorl section, whereas *Rieppelites* acquires a rounded venter. *Rieppelites* is also similar to *Paraceratites* (type species *Ceratites elegans* Mojsisovics, 1882), from which it differs by having a low-arched venter without a keel.

Occurrence. – *Cordeyi* Subzone, *Weitschati* Zone (Late Anisian).

Rieppelites boletzkyi n. sp.

Figs. 33–36; Pl. 22; Pl. 23: 1–8

Diagnosis. – *Rieppelites* with weak ribbing and lateral tuberculation at or just below mid-flank at all growth stages.

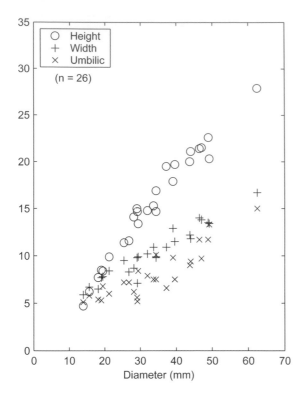
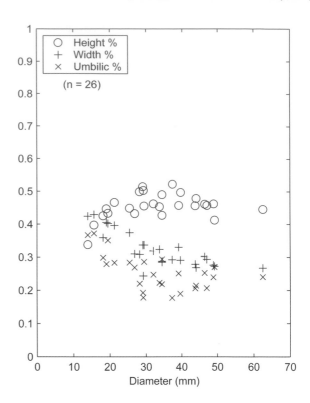

Fig. 32. Scatter diagram of H, W, and U, and of H/D, W/D, and U/D for *Jenksites flexicostatus* n. gen. n. sp. (Augusta Mountains; *Cordeyi* Subzone, *Weitschati* Zone, Late Anisian).

Holotype. – PIMUZ 25372, Loc. HB 2006, Rieber Gulch (Augusta Mountains); *Cordeyi* Subzone, *Weitschati* Zone, Late Anisian.

Etymology. – Species named after S. Boletzky (Banyuls).

Description. – Moderately involute shell with egressive coiling at maturity. Whorl section compressed, high whorled, subrectangular to subhexagonal with a narrow, low-arched venter becoming more rounded at maturity, angular ventral shoulders, and a low umbilical wall. Ornamentation consists of thin, sparse, rectiradiate, sinuous ribs branching at or just below mid-flank. Ribs are projected on ventral shoulders but do not cross the venter. Lateral and ventrolateral rows of tubercles disappear at maturity. Highly variable species ranging from smooth to heavily bituberculate shells. Parabolic nodes restricted to innermost whorls. Suture line ceratitic, with broad saddles.

Measurements. – See Figures 33, 34, and 36. Aside from the innermost whorls, H, W, and U have a normal distribution and they exhibit allometric growth. *Rieppelites boletzkyi* and *R. shevyrevi* have nearly identical growth curves. However, the umbilical diameter is always smaller in *R. boletzkyi*.

Discussion. – *Rieppelites boletzkyi* differs from *R. shevyrevi* by having weaker, more widely spaced ribs, lateral tubercles just below mid-flank of inner whorls, and slightly higher whorls.

Occurrence. – Rieber Gulch (Augusta Mountains): HB 2006 (175), HB 2009 (8), HB 2010 (91), HB 2050 (15), HB 2071 (1); *Cordeyi* Subzone, *Weitschati* Zone (Late Anisian). Oliver Gulch (Augusta Mountains): HB 713 (7); *Cordeyi* Subzone, *Weitschati* Zone (Late Anisian). Muller Canyon (Augusta Mountains): HB 730 (4); *Cordeyi* Subzone, *Weitschati* Zone (Late Anisian). Ferguson Canyon (Augusta Mountains): HB 751 (3); *Cordeyi* Subzone, *Weitschati* Zone (Late Anisian).

Rieppelites shevyrevi n. sp.

Figs. 33, 34, 37, 38; Pls. 20, 21

Diagnosis. – *Rieppelites* with coarse, dense ribbing and lateral tuberculation low on flanks of inner whorls.

Holotype. – PIMUZ 25405, Loc. HB 2030, Ferguson West (Augusta Mountains); *Cordeyi* Subzone, *Weitschati* Zone, Late Anisian.

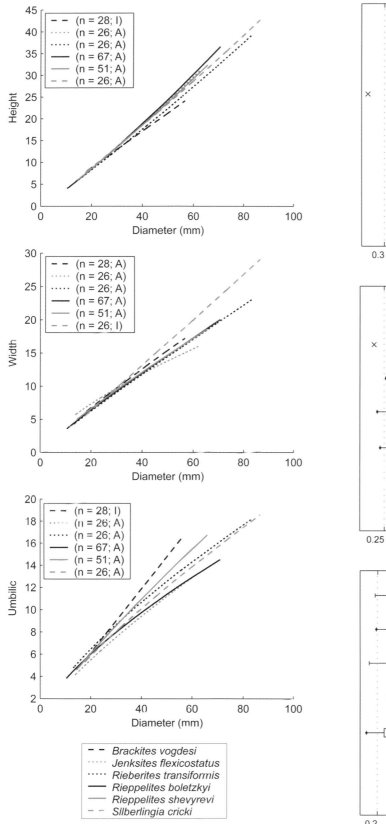

Fig. 33. Growth curves for the studied species of Paraceratinae. I: isometric growth; A: allometric growth.

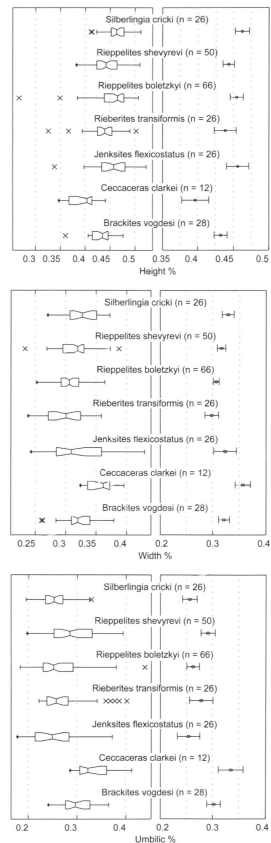

Fig. 34. Box and mean plots for the studied species of Paraceratinae. See text for explanations.

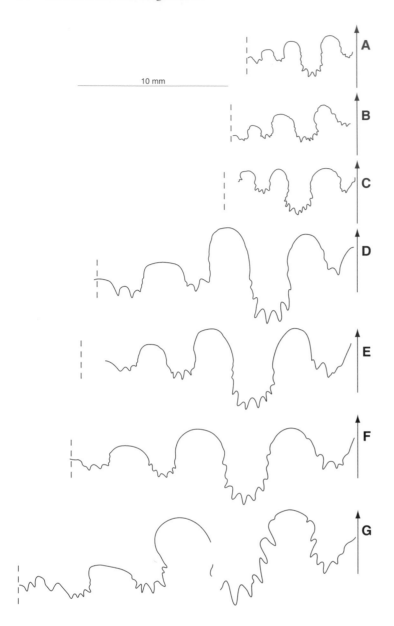

Fig. 35. Suture lines (× 4) of *Rieppelites boletzkyi* n. gen. n. sp. A: Paratype PIMUZ 25380, Loc. HB 2006, Rieber Gulch (Augusta Mountains); *Cordeyi* Subzone, *Weitschati* Zone, Late Anisian. B: Paratype PIMUZ 25381, Loc. HB 2006, Rieber Gulch (Augusta Mountains); *Cordeyi* Subzone, *Weitschati* Zone, Late Anisian. C: Paratype PIMUZ 25388, Loc. HB 2006, Rieber Gulch (Augusta Mountains); *Cordeyi* Subzone, *Weitschati* Zone, Late Anisian; reversed. D: Paratype PIMUZ 25373, Loc. HB 2006, Rieber Gulch (Augusta Mountains); *Cordeyi* Subzone, *Weitschati* Zone, Late Anisian; reversed. E: Paratype PIMUZ 25391, Loc. HB 2006, Rieber Gulch (Augusta Mountains); *Cordeyi* Subzone, *Weitschati* Zone, Late Anisian; reversed. F: Paratype PIMUZ 25374, Loc. HB 2006, Rieber Gulch (Augusta Mountains); *Cordeyi* Subzone, *Weitschati* Zone, Late Anisian; reversed. G: Holotype PIMUZ 25372, Loc. HB 2006, Rieber Gulch (Augusta Mountains); *Cordeyi* Subzone, *Weitschati* Zone, Late Anisian.

Etymology. – Species named after A. A. Shevyrev (Moscow).

Description. – Moderately involute, compressed, high-whorled shell with a subrectangular whorl section, angular ventral and umbilical shoulders, and a low-arched, smooth venter. Ornamentation consisting of dense, sinuous, rectiradiate ribs projected on ventral shoulders but not crossing the venter. Shell bituberculate, with a marginal row of tubercles, and a lateral row low on flanks, which may disappear at maturity. Ribs branching from lateral nodes. Non-tuberculate intercalatories may occur between branching ribs. Ratio of marginal to lateral tubercles is 2:1. Inner whorls with parabolic nodes. One specimen (Pl. 21: 13) clearly shows the transition from parabolic nodes high on flanks to lateral tubercles low on flanks. Suture line ceratitic, with four broad saddles.

Measurements. – See Figures 33, 34, and 38. H, W, and U are highly variable but have a normal distribution throughout ontogeny. All growth curves are allometric and very similar to those of *Rieppelites boletzkyi*. However, *R. shevyrevi* has a higher mean value of U than that of *R. boletzkyi*.

Discussion. – *Rieppelites shevyrevi* differs from *R. boletzkyi* by having denser ribbing, which is predominant over tuberculation, and a lateral row of tubercles low on the flanks. Weakly ornamented variants of *R. shevyrevi* remain ribbed, while those of *R. boletzkyi* are completely smooth.

Occurrence. – Oliver Gulch (Augusta Mountains): HB 592 (3), HB 593 (3), HB 598 (2), HB 713 (2); *Cordeyi* Subzone, *Weitschati* Zone (Late Anisian). Ferguson Canyon (Augusta Mountains): HB 2020 (2?), HB 2030

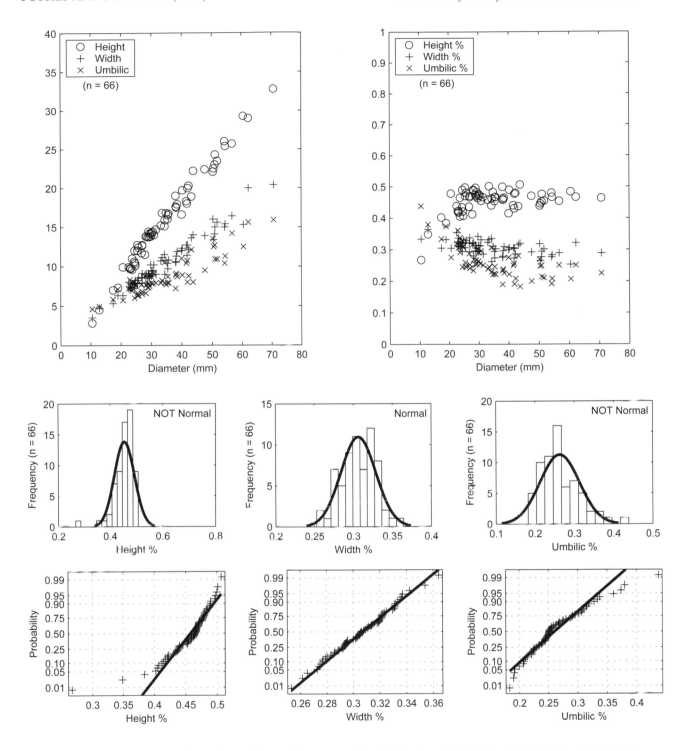

Fig. 36.   Scatter diagram of H, W, and U, and scatter diagram, histograms, and probability plots of H/D, W/D, and U/D for *Rieppelites boletzkyi* n. gen. n. sp. (Augusta Mountains; *Cordeyi* Subzone, *Weitschati* Zone, Late Anisian).

(32), HB 2036 (1?); *Cordeyi* Subzone, *Weitschati* Zone (Late Anisian). Muller Canyon (Augusta Mountains): HB 730 (5), HB 731 (2), HB 734 (41), HB 737 (1); *Cordeyi* Subzone, *Weitschati* Zone (Late Anisian). Rieber Gulch (Augusta Mountains): HB 2010 (7), HB 2012 (1);

Weitschati Zone (Late Anisian). Favret Canyon (Augusta Mountains): JJ 1997/2 (3); *Cordeyi* Subzone, *Weitschati* Zone (Late Anisian). McCoy Mine (New Pass Range): HB 2062 (6), HB 2064 (3); *Cordeyi* Subzone, *Weitschati* Zone (Late Anisian).

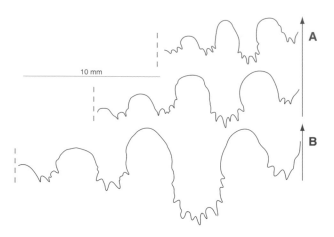

Fig. 37. Suture lines (× 4) of *Rieppelites shevyrevi* n. gen. n. sp. A: PIMUZ 25402, Loc. JJ 1997/2, Favret Canyon (Augusta Mountains); *Cordeyi* Subzone, *Weitschati* Zone, Late Anisian; reversed. B: PIMUZ 25422, Loc. HB 2062, McCoy Mine (New Pass Range); *Cordeyi* Subzone, *Weitschati* Zone, Late Anisian.

Genus *Rieberites* n. gen.

Type species. – *Rieberites transiformis* n. sp.

Diagnosis. – Genus characterized by three successive ontogenetic stages: (i) involute, high-whorled inner whorls with a low-arched, smooth venter, and slightly sinuous ribs swollen below mid-flank; (ii) moderately involute, high-whorled outer whorls with a smooth, subtabulate venter, dense, biconcave ribs swollen on mid-flank, a weak, low-rounded keel, and with or without lateral and marginal tubercles; and (iii) evolute mature whorls with a rounded venter and dense, sinuous ribs swollen on mid-flank.

Etymology. – Genus named after H. Rieber (Zürich).

Composition of the genus. – Type species only.

Description. – As for the type species.

Discussion. – *Rieberites* differs from *Rieppelites*, to which it is closely related, by having biconcave rather than sinuous ribs, a weak keel on outer whorls, and a tabulate venter. *Rieberites* differs from *Jenksites* by having lateral tubercles and a weak keel on outer whorls, flexuous rather than biconcave ribs, and a tabulate venter. *Rieberites* differs from *Silberlingia* by having a weak keel only on outer whorls, biconcave rather than sinuous ribs, and a tabulate venter.

Occurrence. – As for the type species.

Rieberites transiformis n. sp.

Figs. 33, 34, 39; Pl. 24; Pl. 25: 1–2

Holotype. – PIMUZ 25366, Loc. HB 596, Oliver Gulch (Augusta Mountains); *Transiformis* Subzone, *Weitschati* Zone, Late Anisian.

Etymology. – Species name refers to its pronounced ontogenetic changes.

Description. – Inner whorls involute and high whorled, with flat flanks converging towards a low-arched, smooth venter. Whorl section subhexagonal. Dense, sinuous ribs swollen on lower flanks, where tiny tubercles may develop on robust variants.

Outer whorls moderately involute, high whorled, with a subrectangular to subhexagonal whorl section, a smooth, subtabulate venter, angular ventral and umbilical shoulders, and dense, biconcave, rectiradiate to prorsiradiate ribs swollen on mid-flank, where tubercles may develop on robust variants. Ribs projected on ventral shoulders. Clavate tubercles may occur on ventral shoulders of robust variants. Robust variants also bear a faint, low-rounded keel.

Markedly egressive coiling of mature whorls. Mature shell with a subrectangular whorl section, a smooth, low-rounded venter, convex flanks, and dense, sinuous ribs swollen on mid-flank. Tuberculation disappearing at maturity. Suture line unknown.

Measurements. – See Figures 33, 34, and 39. H, W, and U display allometric growth. Since *Rieberites transiformis* is morphologically similar to *Rieppelites*, it is noteworthy that the whorl width growth curve for *Rieberites transiformis* is similar to those of the two species of *Rieppelites*; the growth curve for whorl height is also very similar, but differs slightly by a lower height; the growth curve for umbilical diameter is bracketed between those for the two species of *Rieppelites*.

Discussion. – *Rieberites transiformis* is easily distinguished by its tabulate venter, its biconcave swollen ribs, the presence of a weak keel on outer whorls, and egressive coiling of mature whorls. Otherwise, it shows strong affinities with *Silberlingia praecursor* and robust variants of *Rieppelites boletzkyi*, from which it differs mainly by having a subtabulate venter and biconcave ribs on outer whorls.

Occurrence. – Oliver Gulch (Augusta Mountains): HB 596 (42); *Transiformis* Subzone, *Weitschati* Zone (Late Anisian).

Genus *Silberlingia* n. gen.

Type species. – *Ceratites (Paraceratites) cricki* Smith, 1914

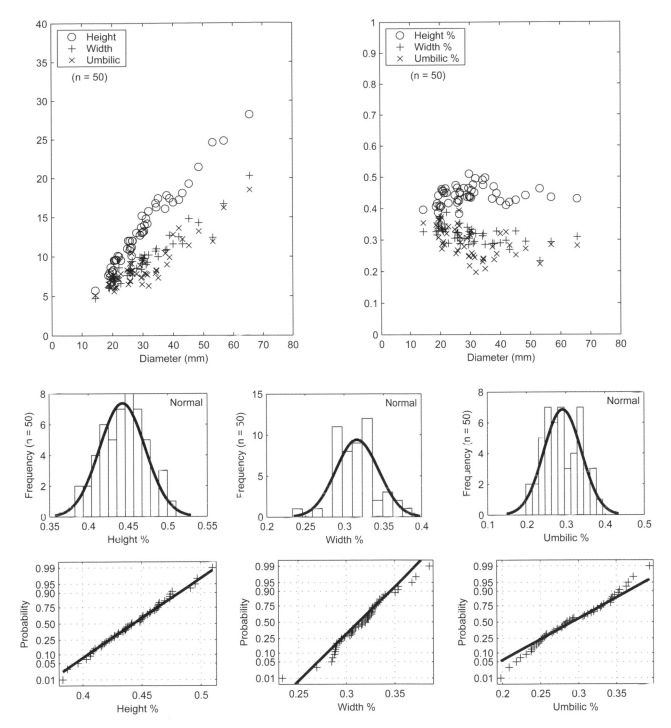

Fig. 38. Scatter diagram of H, W, and U, and scatter diagram, histograms, and probability plots of H/D, W/D, and U/D for *Rieppelites shevyrevi* n. gen. n. sp. (Augusta Mountains; *Cordeyi* Subzone, *Weitschati* Zone, Late Anisian).

Diagnosis. – Moderately involute, high-whorled, compressed shell with a low-arched, smooth, venter, a conspicuous keel extending above ventral shoulders, umbilical tubercles, lateral spines, and marginal clavi. Suture line ceratitic.

Etymology. – Genus named after N. J. Silberling (Lakewood, Co.).

Composition of the genus. – *Ceratites (Paraceratites) cricki* Smith, 1914, *Ceratites (Paraceratites) clarkei* Smith, 1914, and *Silberlingia praecursor* n. sp.

Discussion. – Silberling & Nichols (1982) recognized four species assigned to "*Paraceratites*" (*Ceratites (Paraceratites) cricki* Smith, 1914; *Ceratites (Paraceratites) burckhardti* Smith, 1914; *Ceratites (Paraceratites) clarkei*

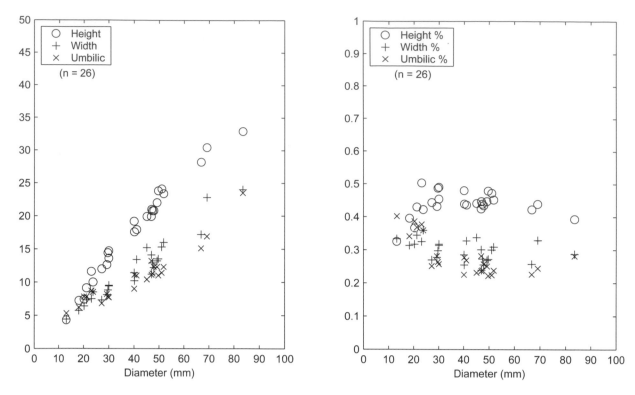

Fig. 39. Scatter diagram of H, W, and U, and of H/D, W/D, and U/D for *Rieberites transiformis* n. gen. n. sp. (Augusta Mountains; *Transiformis* Subzone, *Weitschati* Zone, Late Anisian).

Smith, 1914; and *Ceratites vogdesi* Smith, 1904). However, assignment of these four species to *Paraceratites* is herein rejected. Indeed, the outer whorls of *C. cricki* and *C. burckhardti* differ significantly from those of *C. elegans* Mojsisovics, 1882 (type species of *Paraceratites*) by having coarse lateral tubercles, a more depressed whorl section, and a broad, rounded venter. *C. vogdesi* differs from *C. elegans* by the absence of a keel, by the presence of lateral spines high on flanks, and by more evolute coiling. Only *C. clarkei* displays some affinities with *C. elegans*, from which it nevertheless differs by having a sharper keel, stronger tubercles, more distant ribs, a lateral row of tubercles well below mid-flank, and a less indented suture line. Hence, *Paraceratites sensu* Silberling & Nichols (1982) must be removed from the group of *Paraceratites elegans* Mojsisovics, 1882. *Ceratites (Paraceratites) cricki* Smith, 1914, *Ceratites (Paraceratites) burckhardti* Smith, 1914, and *Ceratites (Paraceratites) clarkei* Smith, 1914 are herein assigned to the new genus *Silberlingia*, and *Ceratites vogdesi* Smith, 1904 to the new genus *Brackites*.

Silberlingia differs from *Brackites* by having a keel, lateral tubercles never above mid-flank, more involute coiling, weaker marginal tubercles that do not develop spines, and higher whorls. *Silberlingia* differs from *Ceccaceras* by having more involute coiling, higher whorls, a more compressed whorl section, and the persistence of the keel on outer whorls. *Silberlingia* differs from

Rieberites by having more widely spaced and sinuous rather than biconcave ribs, a more rounded venter, more convex flanks, higher whorls, and a keel that persists on outer whorls.

Occurrence. – *Weitschati* and *Rotelliformis* zones (Late Anisian).

Silberlingia cricki (Smith, 1914)

Figs. 33, 34, 40; Pl. 25: 5; Pls. 26, 27

1914   *Ceratites (Paraceratites) cricki* – Smith, p. 87; Pl. 37: 6–13; Pl. 38: 1–8; Pl. 47: 19–21.

? 1914   *Ceratites (Paraceratites) cricki* – Smith, p. 87; Pl. 38: 9–12; Pl. 47: 22–24.

1914   *Ceratites (Paraceratites) taurus* – Smith, p. 88; Pl. 35: 1–3.

1914   *Ceratites (Paraceratites) newberryi* – Smith, p. 92; Pl. 40: 6–11.

1914   *Ceratites (Paraceratites) burckhardti* – Smith, p. 90; Pl. 52: 19–21 [holotype].

1982   *Paraceratites cricki* – Silberling & Nichols, p. 34; Pl. 17: 6–20.

1982   *Paraceratites burckhardti* – Silberling & Nichols, p. 31; Pl. 15: 6–18.

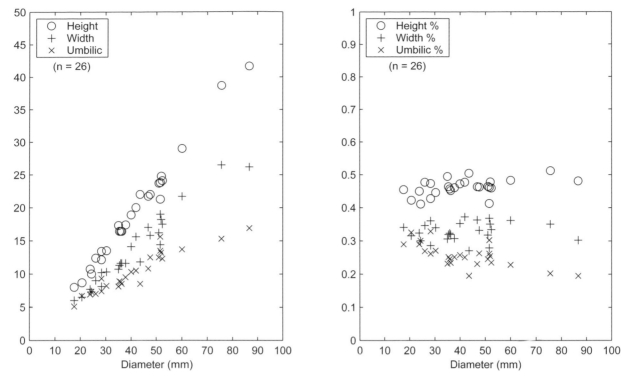

Fig. 40. Scatter diagram of H, W, and U, and of H/D, W/D, and U/D for *Silberlingia cricki* (Smith, 1914) (Augusta Mountains; *Vogdesi* Subzone, *Rotelliformis* Zone, Late Anisian).

Description. – Moderately involute, high-whorled, trituberculate (umbilical, lateral, marginal), and keeled shell. Innermost whorls with a high oval whorl section, a rounded venter, and umbilical bullae only. At a diameter of about 10 mm, the umbilical bullae are replaced by umbilical and lateral rows of tubercles. At a diameter of about 20 mm, a keel and marginal tubercles arise.

Inner whorls with a high-whorled, polygonal to subrectangular, compressed whorl section. Inner whorls trituberculate (umbilical, lateral low on flanks, and marginal), with dense, thin, straight, prorsiradiate ribs, and a low-arched venter bearing a conspicuous keel. Outer whorls similar but with the lateral row of tubercles shifting to mid-flank and enlarging into more widely spaced small spines. Marginal tubercles may alternate on each side of the venter. Marginal and umbilical tubercles lost at maturity, together with a slight egressive coiling.

The shape of the venter varies with the strength of ornamentation, from low-arched with a broad, rounded keel to acute with a high, thinner keel. The ratio of marginal to lateral tubercles also varies with the strength of ornamentation from 3:1 to 2:1. On weakly ornamented variants, the marginal and umbilical rows of tubercles may disappear.

Measurements. – See Figures 33, 34, and 40. H and U have allometric growth, while W has isometric growth.

Discussion. – Silberling & Nichols (1982) followed Smith (1914) and retained the separation of *Ceratites cricki* and

Ceratites burckhardti, although they were certainly mindful of their similarities. According to Silberling & Nichols (1982), *C. cricki* differs from *C. burckhardti* by having a more abrupt and angular marginal edge and a lower ratio of marginal to lateral tubercles. The two species also characterize two non-successive subzones. However, our new collections from Fossil Hill clearly demonstrate that these minor differences fall within intraspecific variability, as documented by samples obtained from single beds.

Silberlingia cricki differs from *S. praecursor* by having umbilical tubercles, weaker, more widely spaced ribs, and stronger tuberculation. *S. cricki* differs from *S. clarkei* by the mid-flank rather than near-umbilical position of lateral tubercles on outer whorls.

Occurrence. – Oliver Gulch (Augusta Mountains): HB 702 (7), HB 746 (2), HB 747 (5), HB 2043 (1); *Rotelliformis* Zone (Late Anisian). Muller Canyon (Augusta Mountains): HB 716 (1); *Rotelliformis* Zone (Late Anisian). Fossil Hill (Humboldt Range): FHB 13 (4), FHB 14 (1?), FHB 15A (20), FHB 17 (2), FHB 22 (9); *Rotelliformis* Zone (Late Anisian).

Silberlingia clarkei (Smith, 1914)

Pl. 9: 6

1914   *Ceratites (Paraceratites) clarkei* – Smith, p. 91; Pl. 40: 15–23; Pl. 52: 1–11.

? 1914   *Ceratites (Paraceratites) trinodosus* – Smith,
          p. 92; Pl. 39: 6; Pl. 52: 15–18.
1914   *Ceratites (Paraceratites) wardi* – Smith, p. 94;
          Pl. 53: 4–6 [holotype], 7–8.
? 1914   *Ceratites beecheri* – Smith, p. 94; Pl. 43: 18–20.
? 1914   *Nevadites fontainei* – Smith, p. 122; Pl. 41:
          16–22; Pl. 51: 5–9.
1982   *Paraceratites clarkei* – Silberling & Nichols,
          p. 32; Pl. 16: 3–4.
? 1982   *Paraceratites clarkei* – Silberling & Nichols,
          p. 32; Pl. 16: 1–2.
non 1982 *Paraceratites clarkei* – Silberling & Nichols,
          p. 32; Pl. 16: 5–12 [= *Ceccaceras stecki* n. gen. n.
          sp.].

Description. – Moderately involute, high whorled, compressed, with trituberculate ornamentation (umbilical, lateral, marginal). Whorl section polygonal to subrectangular. Low-arched venter with a low, rounded keel. Dense, slightly sinuous, prorsiradiate ribs commonly branch from lateral tubercles positioned on lower flanks.

Discussion. – Although the definition of *S. clarkei* by Silberling & Nichols (1982) encompasses *Ceccaceras stecki* n. sp. as defined below, *S. clarkei* is still a valid species closely related to *S. cricki*. *S. clarkei* differs from *S. cricki* by the persistence of umbilical tubercles low on flanks on outer whorls, and it does not develop coarse lateral tubercles on outer whorls. *S. clarkei* differs from *S. praecursor* by having lateral tubercles placed well below mid-flank, and weaker, more widely spaced ribs.

Occurrence. – Muller Canyon (Augusta Mountains): HB 736 (1); Late Anisian. This rare species is still unknown from other controlled stratigraphical successions, and its range is therefore poorly known.

Silberlingia praecursor n. sp.

Fig. 41; Pl. 25: 3–4

Diagnosis. – *Silberlingia* characterized by dense, sinuous ribs, compressed shell, and the absence of umbilical tubercles.

Holotype. – PIMUZ 25449, Loc. HB 596, Oliver Gulch (Augusta Mountains); *Transiformis* Subzone, *Weitschati* Zone, Late Anisian.

Etymology. – Species name refers to the oldest occurrence of the genus.

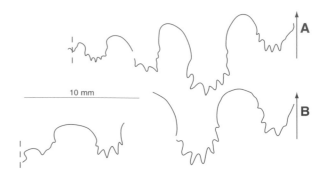

Fig. 41. Suture lines (× 3) of *Silberlingia praecursor* n. gen. n. sp. A: Paratype PIMUZ 25450, Loc. HB 596, Oliver Gulch (Augusta Mountains); *Transiformis* Subzone, *Weitschati* Zone, Late Anisian; reversed. B: Paratype PIMUZ 25451, Loc. HB 596, Oliver Gulch (Augusta Mountains); *Transiformis* Subzone, *Weitschati* Zone, Late Anisian.

Description. – Moderately involute, compressed, high-whorled species with convex flanks, and a narrow, high-arched venter bearing a low, broad keel. Ornamentation consisting of slightly sinuous, prorsiradiate ribs swollen on mid-flank, where they develop into tubercles. Ribs projected on ventral shoulders with clavate tubercles. The ratio between marginal and lateral tubercles is 3:2. Inner whorls unknown. Suture line ceratitic, with four well-rounded saddles.

Measurements. – See Appendix.

Discussion. – *Silberlingia praecursor* differs from the overlying *S. cricki* by having a more compressed whorl section, stronger, denser ribs, and by the absence of umbilical tubercles and spinose lateral tubercles. *Silberlingia praecursor* differs from *S. clarkei* by having lateral tubercles on mid-flank, stronger, denser ribs, and higher whorls.

S. praecursor differs from *Rieberites transiformis*, to which it is closely related, by having sinuous rather than biconcave ribs, more involute coiling, a more rounded venter, and the persistence of a broad, high keel on outer whorls. *S. praecursor* is an unusual species, which combines ornamentation that is intermediate between *R. transiformis* and *S. cricki*. However, its characteristic whorl section and the persistence of a high keel on its outer whorls make its assignment to *Silberlingia* more likely.

Occurrence. – Oliver Gulch (Augusta Mountains): HB 596 (4); *Transiformis* Subzone, *Weitschati* Zone (Late Anisian).

Genus *Ceccaceras* n. gen.

Type species. – *Ceccaceras stecki* n. sp.

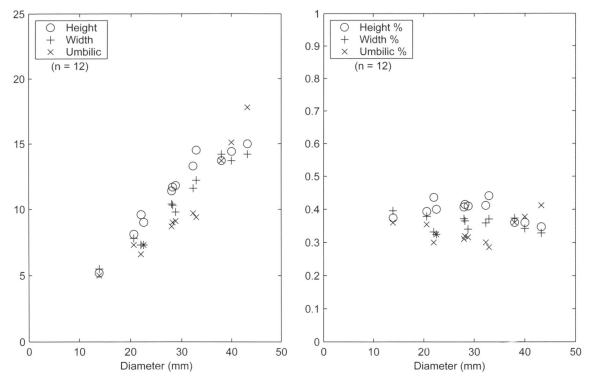

Fig. 42. Scatter diagram of H, W, and U, and of H/D, W/D, and U/D for *Ceccaceras stecki* n. gen. n. sp. (Augusta Mountains; *Vogdesi* Subzone, *Rotelliformis* Zone, Late Anisian).

Etymology. – Genus named after F. Cecca (Paris).

Composition of the genus. – Type species only.

Description. – As for the type species.

Discussion. – Silberling & Nichols (1982) included in *Ceratites (Paraceratites) clarkei* Smith, 1914 forms clearly belonging to a different group. As described by Smith (1914), *C. clarkei* obviously should be referred to *Silberlingia*, which retains a keel and lateral tubercles positioned low on the flanks of the outer whorls. Most specimens figured by Silberling & Nichols (1982) differ in that the keel disappears on the outer whorls, the outer whorls develop a lower, more depressed section, and coiling is moderately evolute. These differences are herein considered to be sufficient justification to erect these specimens to a new genus and species. Furthermore, our new collections confirm the existence of this clearly separate group.

    Ceccaceras differs from *Silberlingia* by having more evolute coiling, lower, more depressed whorls, a quadrate section on outer whorls, and by the disappearance of the keel on outer whorls. *Ceccaceras* differs from *Brackites* by the presence of a distinct keel on inner whorls, the lateral row of tubercles positioned below mid-flank, and a quadrate outer whorl section.

Occurrence. – As for the type species.

Ceccaceras stecki n. sp.

Figs. 34, 42; Pl. 9: 7–13

1982 *Paraceratites clarkei* – Silberling & Nichols, p. 32 [part]; Pl. 16: 5–12

Holotype. – PIMUZ 25207, Loc. FHB 18, Fossil Hill (Humboldt Range); *Vogdesi* Subzone, *Rotelliformis* Zone, Late Anisian.

Etymology. – Species named after A. Steck (Lausanne).

Description. – Inner whorls moderately involute, moderately compressed, with umbilical bullae and thin, straight ribs bearing lateral tubercles low on flanks, and with a low-arched, keeled venter. Outer whorls with a quadrate section, straight, rectiradiate, widely spaced ribs, trituberculate (weak umbilical, strong lateral low on flanks, and strong marginal), and a smooth venter without a keel. Ratio of marginal to lateral tubercles less than 2:1. Umbilical tubercles disappear on mature body chamber. Suture line unknown.

Measurements. – See Figures 34 and 42. Although few specimens are available, this species seems to have a significantly lower mean H and a significantly higher mean W than other studied species of paraceratitins. In other words, this species differs by having a more quadrate whorl section than all other paraceratitins studied herein.

Discussion. – Although known only from a limited amount of specimens, this species is easily distinguishable by its trituberculate and keeled inner whorls, and quadrate mature whorl section without a keel.

Occurrence. – Fossil Hill (Humboldt Range): FHB 15 (1), FHB 18 (15); *Vogdesi* Subzone, *Rotelliformis* Zone (Late Anisian).

Genus *Brackites* n. gen.

Type species. – *Ceratites vogdesi* Smith, 1904

Diagnosis. – Moderately evolute shell with a compressed, high, subrectangular whorl section, umbilical tubercles, lateral spines above mid-flank, small marginal spines alternating on each side of the venter, and a non-keeled, low-arched, smooth venter.

Etymology. – Genus named after P. Brack (Zürich).

Composition of the genus. – *Ceratites vogdesi* Smith, 1904 and *Brackites spinosus* n. sp.

Discussion. – *Ceratites vogdesi* was included in the genus "*Paraceratites*" by Silberling & Nichols (1982). However, they also recognized that the four species they assigned to "*Paraceratites*" are not homogeneous with respect to "*Paraceratites*" *cricki* (here assigned to the new genus *Silberlingia*). Indeed, *Ceratites vogdesi* differs from *Silberlingia* by the absence of a keel, by having a rounded venter not extending above marginal tubercles, a lateral row of tubercles well above mid-flank, a latero-umbilical row of tubercles rather than strictly umbilical, and a more evolute coiling. These differences are believed to be sufficient justification to assign *Ceratites vogdesi* to the new genus *Brackites*.

Brackites differs from *Ceccaceras* by the absence of a keel throughout its entire ontogeny, by having higher outer whorls, and lateral tubercles well above mid-flank on outer whorls. *Brackites* differs from *Nevadites* by having more involute coiling, higher whorls, and a more compressed whorl section.

Occurrence. – *Vogdesi* Subzone, *Rotelliformis* Zone (Late Anisian).

Brackites vogdesi (**Smith, 1904**)

Figs. 33, 34, 43; Pl. 28; Pl. 29: 1–10

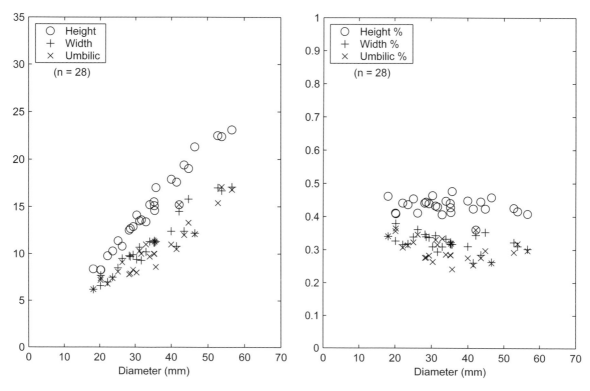

Fig. 43.   Scatter diagram of H, W, and U, and of H/D, W/D, and U/D for *Brackites vogdesi* (Smith, 1904) (Augusta Mountains; *Vogdesi* Subzone, *Rotelliformis* Zone, Late Anisian).

1904 *Ceratites vogdesi* – Smith, p. 384; Pl. 43: 7, 8; Pl. 44: 1.

1914 *Ceratites (Paraceratites) vogdesi* – Smith, p. 89; Pl. 35: 4–6.

? 1914 *Ceratites (Paraceratites) vogdesi* – Smith, p. 89; Pl. 35: 7–9.

1914 *Ceratites (Paraceratites) newberryi* – Smith, p. 92; Pl. 40: 1–5.

? 1914 *Ceratites (Paraceratites) newberryi* – Smith, p. 92; Pl. 40: 13–14.

1914 *Ceratites (Paraceratites) trinodosus* – Smith, p. 92; Pl. 39: 1–5, 9–10; Pl. 52: 12–14.

? 1914 *Ceratites (Paraceratites) trinodosus* – Smith, p. 92; Pl. 39: 7–8, 11–19.

1914 *Ceratites fissicostatus* – Smith, p. 96; Pl. 53: 1–3.

1914 *Ceratites haguei* – Smith, p. 97; Pl. 42: 1–2 [holotype], 3–5; Pl. 43: 1–2a.

? 1914 *Ceratites haguei* – Smith, p. 97; Pl. 43: 3–10.

? 1914 *Ceratites kingi* – Smith, p. 85; Pl. 41: 1–3a [holotype], 4–13.

? 1914 *Ceratites (Paraceratites) trojanus* – Smith, p. 88; Pl. 36: 1–3 [holotype], 4–5; Pl. 37: 1–5.

? 1914 *Ceratites crassicornu* – Smith, p. 95; Pl. 43: figs. 11–12 [holotype], 13–14.

1982 *Paraceratites vogdesi* – Silberling & Nichols, p. 33; Pl. 16: 13–28; Pl. 17: figs. 1–5.

Description. – Inner whorls moderately involute, compressed, and high, with a subrectangular whorl section, moderately angular ventral shoulders, and a smooth, low-arched venter without a keel. Ornamentation consisting of dense, thin, slightly prorsiradiate, straight to slightly sinuous ribs, umbilical tubercles, lateral spines high on flanks, and marginal tubercles alternating on each side of the venter. Outer whorls becoming moderately evolute, with umbilical tubercles shifting to a latero-umbilical position and changing into slightly projected bullae, and enlarged lateral spines being more widely spaced. The ratio between marginal and lateral tubercles is 2:1 for inner whorls and 3:1 for outer whorls. Umbilical tubercles may disappear at maturity. Ribbing decreasing in strength throughout ontogeny, with a tendency to fade on outer whorls.

Measurements. – See Figures 33, 34, and 43. Whorl height and umbilical diameter display allometric growth, whereas whorl width displays isometric growth.

Discussion. – *Brackites vogdesi* differs from *B. spinosus* by having a less spinose sculpture, more involute coiling, a more subrectangular whorl section, and by the absence of a second row of umbilical tubercles.

Occurrence. – Oliver Gulch (Augusta Mountains): HB 742 (15), HB 745 (19), HB 746 (2), HB 749 (2), HB 2042 (1?);

Rotelliformis Zone (Late Anisian). Muller Canyon (Augusta Mountains): HB 736 (2); *Rotelliformis* Zone (Late Anisian). Favret Canyon (Augusta Mountains): HB 2047 (6), HR ##B (8); *Rotelliformis* Zone (Late Anisian). Ferguson Canyon (Augusta Mountains): HB 2023 (8); *Rotelliformis* Zone (Late Anisian). McCoy Mine (New Pass Range): HB 2063 (4); *Rotelliformis* Zone (Late Anisian). Fossil Hill (Humboldt Range): FHB 15 (1), FHB 15A (10), FHB 20 (1); *Rotelliformis* Zone (Late Anisian).

Brackites spinosus n. sp.

Pl. 29: 11–12

Diagnosis. – *Brackites* with a more rounded whorl section, highly spinose tuberculation, a small, weak fourth row of tubercles on umbilical shoulder, and more evolute coiling.

Holotype. – PIMUZ 25183, Loc. HB 703, Oliver Gulch (Augusta Mountains); *Vogdesi* Subzone, *Rotelliformis* Zone, Late Anisian.

Etymology. – Species name refers to the highly spinose tuberculation of this species.

Description. – Very evolute shell with rounded, non-keeled, smooth venter, rounded ventral shoulders, convex upper flanks, flat lower flanks, and narrow umbilical shoulders. Ornamentation consisting of small, umbilical tubercles, latero-umbilical tubercles, lateral spines high on flanks, and small, marginal spines alternating on each side of the venter. Spinose tuberculation appears at a diameter of about 10 mm. Marginal spines slightly outnumbering lateral spines (4:3). Thin, straight, faint ribs developing between umbilical and lateral tubercles. Suture line unknown.

Measurements. – See Appendix.

Discussion. – The quadrituberculate, highly spinose, evolute coiling, and absence of a keel make this species easily distinguishable from other contemporaneous paraceratitins. It differs from *B. vogdesi* by having a small fourth row of umbilical tubercles, a highly spinose sculpture, and a more rounded whorl section. This species displays affinities with the Alpine *Ceratites ecarinatus* Hauer, 1896, which differs essentially by the absence of the fourth umbilical row of tubercles.

Occurrence. – Oliver Gulch (Augusta Mountains): HB 703 (2); *Rotelliformis* Zone (Late Anisian).

Genus *Marcouxites* n. gen.

Type species. – *Ceratites spinifer* Smith, 1914

Diagnosis. – Moderately involute shell with a low-arched, smooth venter, widely spaced, sinuous ribs projected on ventral shoulders, pinched bullae just below mid-flank, clavate marginal tubercles, and oblique, large umbilical shoulders.

Etymology. – Genus named after J. Marcoux (Paris).

Composition of the genus. – Type species only.

Description. – As for the type species.

Discussion. – *Marcouxites* is an unusual genus readily distinguishable by having large, oblique umbilical shoulders. Although superficially similar to *Frechites*, *Marcouxites* is distinguishable by having a more rounded venter, a more compressed section, sparser ribbing, large, oblique umbilical shoulders, and a faint keel. *Marcouxites* differs from *Rieberites* by having a low-arched venter throughout ontogeny, more widely spaced and rather sinuous ribs, lateral bullae rather than simple tubercles, and oblique umbilical shoulders. *Marcouxites* differs from *Silberlingia* by the loss of its keel on outer whorls, by having a more depressed whorl section, more involute coiling, and oblique umbilical shoulders. *Marcouxites* differs from *Ceccaceras* by having higher whorls, more involute coiling, oblique umbilical shoulders, and by the absence of umbilical tubercles. *Marcouxites* differs from *Brackites* by having lateral tubercles below mid-flank, keeled inner whorls, more involute coiling, oblique umbilical shoulders, and by the absence of marginal spines.

Occurrence. – As for the type species.

Marcouxites spinifer (Smith, 1914)

Pl. 9: 4–5

1914 *Ceratites spinifer* – Smith, p. 103; Pl. 59: 1–10; Pl. 60: 1–12.
1934 *Frechites spinifer* – Spath, p. 448.
1982 *Frechites nevadanus* – Silberling & Nichols, p. 29.

Description. – Moderately involute shell with a low-arched, broad, smooth venter, convex flanks, slightly angular ventral shoulders, and oblique, large umbilical shoulders. Ornamentation consists of sparse, sinuous, subdued, rectiradiate ribs branching from pinched bullae just below mid-flank. Ribs projected on ventral shoulders, with slightly clavate marginal tubercles. Inner whorls have low, broad, rounded keel, which disappears on outer

whorls. Suture line not visible on our specimens, but is ceratitic according to Smith (1914).

Measurements. – See Appendix.

Discussion. – Silberling & Nichols (1982) followed the proposal of Spath (1934) who assigned *Ceratites spinifer* Smith, 1914 to *Frechites*. However, *C. spinifer* differs by having two rows of tubercles (marginal and just below mid-flank) and an oblique umbilical wall. Such characteristics are not found on any representative of *Frechites*, and hence the species *C. spinifer* is deserving of separate specific status. Moreover, we also believe that these characteristics justify the erection of *C. spinifer* to a distinct genus, namely *Marcouxites*. Our stratigraphically controlled specimens from the Fossil Hill area come from a distinct level bracketed between the *Lawsoni* and *Vogdesi* subzones (*Rotelliformis* Zone), which is much earlier than the first occurrence of *Frechites*.

Occurrence. – Fossil Hill (Humboldt Range): FHB 11 (2); *Spinifer* Subzone, *Rotelliformis* Zone (Late Anisian).

Genus *Eutomoceras* Hyatt, 1877

Eutomoceras cf. *E. dalli* Smith, 1914

Pl. 30: Fig. 5

1914   *Eutomoceras (Halilucites) dalli* – Smith, p. 65; Pl. 29: 1–4 [holotype], 5–8.
? 1914   *Eutomoceras (Halilucites) dalli* – Smith, p. 65; Pl. 29: 9–10.
1914   *Hungarites fittingensis* – Smith, p. 58 [part]; Pl. 90: 5–7 [holotype].
1982   *Eutomoceras dalli* – Silberling & Nichols, p. 35; Pl. 18: 2–7.
? 1982   *Eutomoceras* cf. *E. lahontanum* – Silberling & Nichols, p. 34; Pl. 18: 1 [not *E. lahontanum*].

Description. – Crushed fragment with high whorl height (H = 28 mm) and prominent papillation on thin ribs arising from umbilical bullae. Last portion of whorl ornamented by numerous, thin, prorsiradiate, slightly sinuous ribs projected forward on ventral shoulders. Acute venter with a thin, high keel.

Discussion. – Our single fragmentary specimen is referred without doubt to *Eutomoceras*, but its specific assignment remains uncertain. The thin, dense ribbing at an estimated diameter of about 65 mm is similar to that of *E. dalli*, as figured by Silberling & Nichols (1982, Pl. 18: 3, 6).

Occurrence. – Fossil Hill (Humboldt Range): FHB 31 (1); *Blakei* Subzone, *Rotelliformis* Zone (Late Anisian).

Eutomoceras dunni Smith, 1904

Pl. 30: 1–4

1904 *Eutomoceras dunni* – Smith, p. 381; Pl. 43: 11 [holotype]; Pl. 44: 4 [holotype].
1914 *Eutomoceras dunni* – Smith, p. 62; Pl. 27: 14–21.
? 1914 *Eutomoceras dunni* – Smith, p. 62; Pl. 27: 22–25.
1914 *Eutomoceras breweri* – Smith, p. 61: Pl. 28: 1–4 [holotype], 5–7a.
1914 *Eutomoceras laubei* – Smith, p. 63; Pl. 26: 7–9.
? 1914 *Hungarites fittingensis* – Smith, p. 58; Pl. 29: 12–14.
1982 *Eutomoceras dunni* – Silberling & Nichols, p. 35; Pl. 18: 8–15.

Description. – Involute, compressed, discoidal, high-whorled shell with slightly angular ventral shoulders, convex to flat flanks converging towards the venter, and an acute, narrow venter bearing a thin, high keel. Ornamentation consisting of dense, sinuous, slightly prorsiradiate ribs, irregularly branching at the umbilicus or with intercalatories on upper flanks. Ribs projected on ventral shoulders. Primary ribs are thicker on lower flanks and arise from strong umbilical bullae. Papillated tuberculation begins at a diameter of about 25 mm.

Measurements. – See Appendix.

Discussion. – Our few specimens agree well with the ornamentation and shell proportion of *E. dunni* as revised by Silberling & Nichols (1982). *E. dunni* differs from coexisting and overlying species mainly by having an intermediate umbilical diameter (U/D about 0.20) and moderate robustness. Specimen PIMUZ 25261 provides the oldest occurrence of the genus and species in Nevada.

Occurrence. – Fossil Hill (Humboldt Range): FHB 12 (1), FHB 34 (5); *Rotelliformis* Zone (Late Anisian).

Superfamily Pinacocerataceae Mojsisovics, 1879

Family Japonitidae Tozer, 1971

Genus *Tropigymnites* Spath, 1951

Tropigymnites sp. indet.

Fig. 44; Pl. 17: 13–14

Description. – Involute, low-whorled, compressed shell with high oval whorl section, a narrow, slightly acute venter, and convex flanks converging towards the venter. Ornamentation consisting of curved to straight,

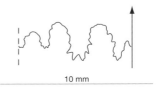

Fig. 44. Suture line (× 4) of *Tropigymnites* sp. indet. PIMUZ 25457, Loc. HB 713, Oliver Gulch (Augusta Mountains); *Cordeyi* Subzone, *Weitschati* Zone, Late Anisian.

prorsiradiate striae, and occasional weak umbilical folds. Suture line subammonitic, composed of slightly indented saddles; first and second saddles about the same size.

Measurements. – See Appendix.

Discussion. – Shape, ornamentation, and suture line of our specimens are similar to those figured by Silberling & Nichols (1982) from the *Vogdesi* Subzone at Fossil Hill. The generic assignment to *Tropigymnites* instead of *Anagymnites*, which has been questioned by several authors (e.g. Spath 1951; Silberling & Nichols 1982), is here confirmed by the subammonitic suture line of some of our specimens. However, our specimens are more involute and have a lower whorl height than the Nevada *T. planorbis* and the Canadian *T. haueri*.

Occurrence. – Oliver Gulch (Augusta Mountains): HB 592 (1), HB 713 (1); *Weitschati* Zone (Late Anisian). Rieber Gulch (Augusta Mountains): HB 2007 (1); *Weitschati* Zone (Late Anisian). McCoy Mine (New Pass Range): HB 2062 (1); *Weitschati* Zone (Late Anisian).

Family Gymnitidae Waagen, 1895

Subfamily Gymnitinae Waagen, 1895

Genus *Gymnites* Mojsisovics, 1882

Gymnites sp. indet.

Pl. 4: 5–7

Description. – Very evolute, highly compressed, high-whorled shell with slightly convex flanks, a highly arched venter, narrow but rounded ventral shoulders, and shallow, convex umbilical wall. Shell smooth, with prorsiradiate, straight striae. Faint, thin, irregular folds may occur on upper flanks and venter. Suture line and body chamber unknown.

Measurements. – See Appendix.

Discussion. – The unknown suture line as well as the absence of mature whorls precludes any firm identification of our specimens at the species level. Five species of *Gymnites* are already known from the Fossil Hill Member (*G. billingsi* and *G. tregorum* of Early Anisian age; *G. tozeri* and *G. perplanus* of Middle Anisian age; *G. humboldti* of Late Anisian age). Our specimens resemble *G. tregorum* by having weak, surficial folds and low whorls, but differ from the latter by having more evolute coiling. Our specimens differ from *G. billingsi* by having lower whorls and more evolute coiling. They also differ from *G. tozeri* by having more evolute coiling, and from *G. perplanus* by having lower whorls and more evolute coiling. *G. humboldti* is only defined by a body chamber, which is completely smooth, and rather involute.

Occurrence. – Ferguson Canyon (Augusta Mountains): HB 739 (5); *Mojsvari* Subzone, *Shoshonensis* Zone (Middle Anisian).

Genus *Anagymnites* Hyatt, 1900

***Anagymnites* sp. indet.**

Fig. 45; Pl. 17: 10–12

Description. – Moderately involute, compressed, high-whorled, discoidal shell with an acute, narrow venter, convex flanks converging towards the venter, rounded ventral shoulders, and a low umbilical wall. Ornamentation consisting of weak, sinuous, rectiradiate, thin ribs. Suture line ammonitic but weakly indented.

Measurements. – See Appendix.

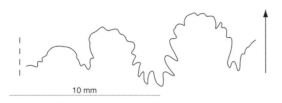

10 mm

Fig. 45. Suture line (× 4) of *Anagymnites* sp. indet. PIMUZ 25131, Loc. FHB 8, Fossil Hill (Humboldt Range); *Lawsoni* Subzone, *Mimetus* Zone, Late Anisian.

Discussion. – The absence of outer whorls precludes firm identification of our specimens at the species level.

Occurrence. – Fossil Hill (Humboldt Range): FHB 6 (1?), FHB 8 (9), FHB 11 (3); *Mimetus* and *Rotelliformis* zones (Late Anisian). Oliver Gulch (Augusta Mountains): HB 596 (1), HB 2043 (1); *Weitschati* Zone (Late Anisian).

Superfamily Ptychitaceae Mojsisovics, 1882

Family Sturiidae Kiparisova, 1958

Genus *Discoptychites* Diener, 1916

***Discoptychites* cf. *D. megalodiscus* (Beyrich, 1867)**

Fig. 46; Pl. 23: 10

1867   *Ammonites megalodiscus* – Beyrich, p. 135; Pl. 2.
1882   *Ptychites megalodiscus* – Mojsisovics, p. 253; Pl. 77: 1a–c; Pl. 78: 1–2.
? 1914   *Ptychites evansi* – Smith, p. 47; Pl. 21: 3–3a.
? 1982   *Discoptychites* sp. – Silberling & Nichols, p. 41; Text-figs. 23–24.

Description. – Very involute (almost closed umbilicus), discoidal, high-whorled shell with a subtriangular whorl section, a narrow, rounded, smooth venter, inconspicuous ventral shoulders, flat flanks converging towards the venter, and high umbilical wall. Phragmocone smooth or with blunt, broad, radial folds. Suture line ammonitic. The largest specimen (not figured) reaches a diameter of 185 mm and is still completely septate.

Measurements. – See Appendix.

Discussion. – Our specimens have the compressed, subtriangular whorl shape and suture pattern that is typical of *Discoptychites*. Although they occur in the *Weitschati* Zone, they are somewhat similar in shape and size to specimens from the *Rotelliformis* Zone described by Silberling & Nichols (1982) and Smith (1914) as *Discoptychites* sp. and *Ptychites evansi*, respectively. Because of insufficient material, Silberling & Nichols

10 mm

Fig. 46. Suture line (× 3) of *Discoptychites* cf. *D. megalodiscus* (Beyrich, 1867). PIMUZ 25218, Loc. HB 596, Oliver Gulch (Augusta Mountains); *Transiformis* Subzone, *Weitschati* Zone, Late Anisian.

(1982) considered *P. evansi* Smith, 1914 as a *nomen dubium*. Our material closely resembles the Alpine *Discoptychites megalodiscus* of the *Trinodosus* Zone.

Occurrence. – Oliver Gulch (Augusta Mountains): HB 591 (1), HB 596 (2); *Weitschati* Zone (Late Anisian). Ferguson Canyon (Augusta Mountains): HB 738 (1); *Cordeyi* Subzone, *Weitschati* Zone (Late Anisian).

Family Ptychitidae Mojsisovics, 1882

Genus *Ptychites* Mojsisovics, 1875

Ptychites sp. indet.

Pl. 23: 9, 11, 12

Description. – Very involute, low-whorled, relatively depressed shell with a discoidal whorl section, convex flanks converging towards a broadly rounded, smooth venter, inconspicuous ventral shoulders, steep umbilical wall, and abrupt umbilical shoulders. Ornamentation consisting of convex, slightly prorsiradiate, thin to thick ribs, fading towards the venter. Innermost whorls bearing varices. Suture line unknown.

Measurements. – See Appendix.

Discussion. – The small specimens available were found at different localities and are quite rare. Among the many species assigned to *Ptychites*, our specimens display affinities with both the *P. megalodisci* and *P. opulenti* groups, but identification at the species level is hampered by insufficient material.
   Our specimens differ from the older species *P. densistriatus* and *P. gradinarui* (*Shoshonensis* Zone) by having a more compressed whorl section, a narrower venter, and more robust sculpture. Furthermore, *Ptychites* sp. indet. differs from *P. densistriatus* by having a more triangular whorl section, and from *P. gradinarui* by having more involute coiling.

Occurrence. – Rieber Gulch (Augusta Mountains): HB 2007 (1); *Cordeyi* Subzone, *Weitschati* Zone (Late Anisian). McCoy Mine (New Pass Range): HB 2062 (4); *Cordeyi* Subzone, *Weitschati* Zone (Late Anisian).

Superfamily Danubitaceae Spath, 1951

Family Longobarditidae Spath, 1951

Subfamily Longobarditinae Spath, 1951

Genus *Longobardites* Mojsisovics, 1882

Longobardites parvus (Smith, 1914)

Figs. 47, 48; Pl. 31: Figs. 1–7

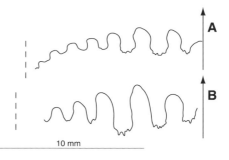

Fig. 47. Suture lines (× 4) of *Longobardites parvus* (Smith, 1914). A: PIMUZ 25344, Loc. HB 2062, McCoy Mine (New Pass Range); *Cordeyi* Subzone, *Weitschati* Zone, Late Anisian; reversed. B: PIMUZ 25342, Loc. HB 713, Oliver Gulch (Augusta Mountains); *Cordeyi* Subzone, *Weitschati* Zone, Late Anisian.

1914   *Dalmatites parvus* – Smith, p. 60; Pl. 30: 1–2 [holotype], 3–7.

? 1914   *Dalmatites parvus* – Smith, p. 60; Pl. 30: 8–12.

1914   *Dalmatites minutus* – Smith, p. 59; Pl. 29: 15–16 [holotype], 17–21.

1982   *Longobardites parvus* – Silberling & Nichols, p. 50; Pl. 21: 19–25; Text-fig. 33.

Description. – Extremely involute (closed umbilicus), oxycone shell with an acute venter but bordered by distinct ventral shoulders, convex flanks, and inconspicuous umbilical shoulders. Innermost whorls have a broader venter. Ornamentation smooth or with convex to sinuous striae or thin folds.

Measurements. – See Figure 48. Height and width of whorl section have narrow variations. H displays isometric growth, whereas W displays allometric growth.

Discussion. – The Nevada longobarditins have been thoroughly revised by Silberling & Nichols (1982), who documented their ontogenetic and intraspecific variability. *Longobardites parvus* differs from the younger *L. zsigmondyi* mainly by having distinct ventral shoulders and a more robust (less compressed) shell shape.

Occurrence. – Oliver Gulch (Augusta Mountains): HB 583 (25), HB 584 (131), HB 585 (24), HB 590(5), HB 597 (1?), HB 708 (4), HB 713 (4); *Weitschati* Zone. Ferguson Canyon (Augusta Mountains): HB 740 (3), HB 2033 (2), HB 2034 (39), HB 2036 (1), HB 2053 (7); *Weitschati* Zone (Late Anisian). Muller Canyon (Augusta Mountains): HB 737 (4); *Weitschati* Zone (Late Anisian). Rieber Gulch (Augusta Mountains): HB 2003 (16), HB 2004 (17), HB 2012 (5), HB 2013 (2), HB 2014 (3), HB 2017 (4); *Weitschati* Zone (Late Anisian). McCoy Mine (New Pass Range): HB 2062 (5); *Weitschati* Zone (Late Anisian). Fossil Hill (Humboldt Range): FHB 9 (1); *Lawsoni* Subzone, *Rotelliformis* Zone (Late Anisian).

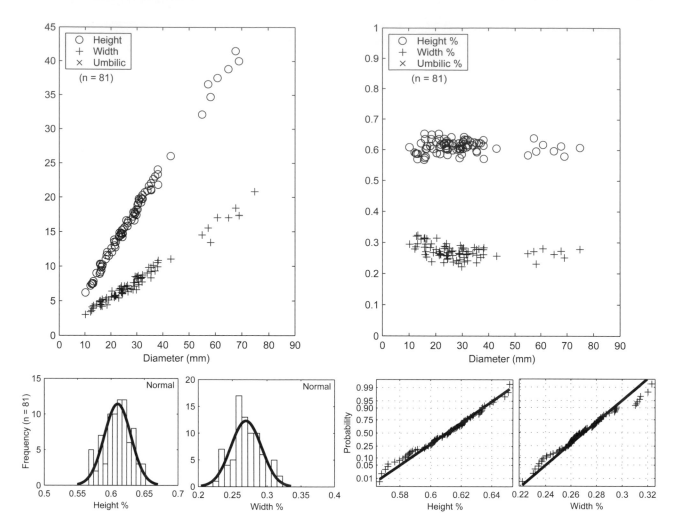

Fig. 48. Scatter diagram of H and W, and scatter diagram, histograms, and probability plots of H/D and W/D for *Longobardites parvus* (Smith, 1914) (Augusta Mountains; Late Anisian).

Longobardites zsigmondyi (Böckh, 1874)

Fig. 49; Pl. 31: 10–13

1874   *Ammonites (Sageceras) zsigmondyi* – Böckh, p. 177; Pl. 4: 14.

1882   *Longobardites zsigmondyi* – Mojsisovics, p. 185; Pl. 52: 4a–c.

1905   *Longobardites nevadanus* – Hyatt & Smith, p. 132 [part]; Pl. 58: 16–18; Pl. 75: 6–7.

? 1905   *Longobardites nevadanus* – Hyatt & Smith, p. 132 [part]; Pl. 58: 1–20; Pl. 75: 8–9.

1914   *Longobardites nevadanus* – Smith, p. 50 [part]; Pl. 8: 16–18; Pl. 12: 6–7; Pl. 30: 13–14.

? 1914   *Longobardites nevadanus* – Smith, p. 50 [part]; Pl. 8: 19–20; Pl. 12: 8–9; Pl. 30: 15–16.

1982   *Longobardites* cf. *zsigmondyi* – Silberling & Nichols, p. 51; Pl. 21: 26–27; Text-fig. 34.

? 1982   *Longobardites* cf. *zsigmondyi* – Silberling & Nichols, p. 51; Pl. 21: 28 [= *Oxylongobardites acutus*?].

Description. – Extremely involute (closed umbilicus), oxycone shell with an acute, narrowly tapered venter, without ventral shoulders, with convex flanks, and inconspicuous umbilical shoulders. Innermost whorls have a broader venter and more conspicuous ventral shoulders. Outer shell smooth or with convex striae.

Measurements. – See Figure 49. As is the case for *Longobardites parvus*, whorl height and width of *L. zsigmondyi* have narrow fluctuations. Both parameters display isometric growth.

Discussion. – Two species of *Longobardites* are recorded from Nevada. The younger *L. zsigmondyi* acquires the mature shell shape and suture pattern of *L. parvus* at a smaller size (Silberling & Nichols 1982). *L. zsigmondyi* differs from *L. parvus* by having a more compressed shell, more involute coiling, and by having an acute venter without distinct ventral shoulders.

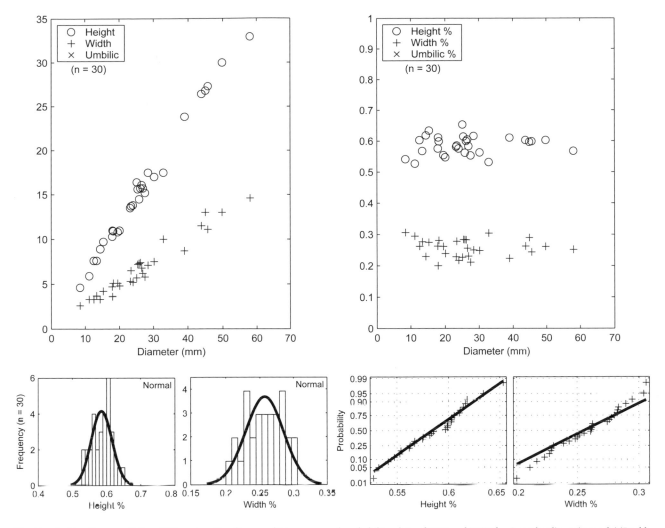

Fig. 49. Scatter diagram of H and W, and scatter diagram, histograms, and probability plots of H/D and W/D for *Longobardites zsigmondyi* (Böckh, 1874) (Augusta Mountains; Late Anisian).

Our specimens record the oldest occurrence of *L. zsigmondyi* in the Fossil Hill Member, which was previously known only from the *Occidentalis* Zone. Absence of *L. zsigmondyi* in the *Meeki* Zone between the *Rotelliformis* and *Occidentalis* zones is probably due to the scarcity of this species. For example, our documented specimens are from the type locality at Fossil Hill but come from the *Lawsoni* Subzone, which apparently was not documented by Silberling & Nichols (1982).

Occurrence. – Oliver Gulch (Augusta Mountains): HB 593 (3), HB 598(1), HB 702 (1), HB 704 (1), HB 707 (1), HB 710 (1), HB 711 (1), HB 713 (1), HB 749 (1); *Weitschati* and *Rotelliformis* zones (Late Anisian). Muller Canyon (Augusta Mountains): HB 735 (15), HB 737 (1); *Weitschati*, *Mimetus*, and *Rotelliformis* zones (Late Anisian). Rieber Gulch (Augusta Mountains): HB 2007 (5); *Weitschati* Zone (Late Anisian). Ferguson Canyon (Augusta Mountains): HB 2022 (2); *Weitschati* Zone (Late Anisian). Fossil Hill (Humboldt Range): FHB 6 (1),

FHB 9 (7); *Lawsoni* Subzone, *Rotelliformis* Zone (Late Anisian).

Genus *Oxylongobardites* n. gen.

Type species. – *Oxylongobardites acutus* n. sp.

Etymology. – Genus name refers to the extreme oxycone shape and the supposed phylogenetic relationship with *Longobardites*.

Composition of the genus. – Type species only.

Description. – As for the type species.

Discussion. – *Oxylongobardites* differs from *Longobardites* by having an extremely acute venter with concave ventral shoulders.

Occurrence. – As for the type species.

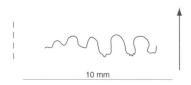

10 mm

Fig. 50.   Suture line (× 4, reversed) of *Oxylongobardites acutus* n. gen. n. sp. Holotype PIMUZ 25352, Loc. HB 596, Oliver Gulch (Augusta Mountains); *Transiformis* Subzone, *Weitschati* Zone, Late Anisian.

Oxylongobardites acutus n. sp.

Fig. 50; Pl. 31: 8–9

? 1982   *Longobardites* cf. *zsigmondyi* – Silberling & Nichols, p. 51; Pl. 21: 28.

Holotype. – PIMUZ 25352, Loc. HB 596, Oliver Gulch (Augusta Mountains); *Transiformis* Subzone, *Weitschati* Zone, Late Anisian.

Etymology. – Species name refers to the very acute shell shape.

Description. – Extremely involute (closed umbilicus), oxycone, highly compressed shell with an acute venter, and convex flanks, slightly concave near venter. Outer shell smooth or with convex striae or thin folds. Adult whorls unknown. Suture line ceratitic, with high, narrow saddles and a pseudo-adventitious longobarditin pattern.

Measurements. – See Appendix.

Discussion. – *Oxylongobardites acutus* is thought to be related to *Longobardites zsigmondyi*. The lineage *L. parvus* to *L. zsigmondyi* to *O. acutus* depicts a trend in whorl section from an acute venter with shoulders to a very acute venter with slightly concave flanks near the venter. Both genera share a similar suture line.

Occurrence. – Oliver Gulch (Augusta Mountains): HB 596 (3); *Transiformis* Subzone, *Weitschati* Zone (Late Anisian). Ferguson West (Augusta Mountains): HB 2030 (1), HB 2036 (1); *Cordeyi* Subzone, *Weitschati* Zone (Late Anisian).

Superfamily Nathorstitaceae (Spath, 1951)

Family Proteusitidae Spath, 1951

Genus *Tropigastrites* Smith, 1914

Tropigastrites lahontanus Smith, 1914

Pl. 30: 11–13

1914   *Tropigastrites lahontanus* – Smith, p. 28; Pl. 19: 14–15a.
1914   *Tropigastrites halli* – Smith, p. 27; Pl. 18: 11–12.
? 1914   *Tropigastrites halli* – Smith, p. 27; Pl. 18: 13–14a; Pl. 88: 14–15.
1914   *Tropigastrites rothpletzi* – Smith, p. 31; Pl. 19: 1–3 [holotype], 4–7; Pl. 8: 24–26.
? 1914   *Tropigastrites rothpletzi* – Smith, p. 31; Pl. 19: 8–13a, 22–23.
1914   *Tropigastrites neumayri* – Smith, p. 29; Pl. 18: 15–16a, 17–17a.
? 1914   *Tropigastrites neumayri* – Smith, p. 29; Pl. 18: 18–21a, 22–23.
? 1914   *Gymnites (Anagymnites) rosenbergi* – Smith, p. 55; Pl. 26: 2–3a [holotype], 4–6.
1982   *Tropigastrites lahontanus* – Silberling & Nichols, p. 54; Pl. 26: 5–17.

Description. – Moderately evolute shell with a subtriangular whorl section. Flanks convex, converging towards the narrowly arched, fastigate venter, and a low umbilical wall. Ornamentation consisting of thin folds or ribs fading on upper flanks; growth striae projected on the venter. Outer whorls becoming more compressed and smoother.

Measurements. – See Appendix.

Discussion. – According to Silberling & Nichols (1982), this species differs from the overlying *Tropigastrites louderbacki* by having thinner, more tightly coiled whorls. Since our specimens come from the *Vogdesi* Subzone, they document the oldest occurrence of the genus in the Fossil Hill Member as compared to those from the *Blakei* Subzone, as recorded by Silberling & Nichols (1982).

Occurrence. – Fossil Hill (Humboldt Range): FHB 17 (1), FHB 30(1), FHB 32 (1?), FHB 34 (1); *Rotelliformis* Zone (Late Anisian).

Tropigastrites louderbacki (Hyatt & Smith, 1905)

Pl. 30: 10

1905   *Sibyllites louderbacki* – Hyatt & Smith, p. 58; Pl. 74: 10–12 [holotype].
1914   *Tropigastrites louderbacki* – Smith, p. 29; Pl. 11: 10–12; Pl. 18: 3–6; Pl. 88: 4–9.
? 1914   *Tropigastrites louderbacki* – Smith, p. 29; Pl. 18: 9–10a; Pl. 88: 10–13.
1914   *Tropigastrites powelli* – Smith, p. 31; Pl. 18: 1–2, 7–8a; Pl. 97: 1–4 [holotype], 5–6, 9–12.
? 1914   *Tropigastrites powelli* – Smith, p. 31; Pl. 97: 7–8.
1914   *Tropigastrites trojanus* – Smith, p. 32; Pl. 17: 1–4 [holotype], 6–9, 11–19.

? 1914 *Tropigastrites trojanus* – Smith, p. 32; Pl. 17: 5, 10, 20–30.

? 1914 *Tropigastrites obliterans* – Smith, p. 32; Pl. 17: 5, 10, 20–30.

1982 *Tropigastrites louderbacki* – Silberling & Nichols, p. 30; Pl. 87: 27–29 [holotype], 30–32.

Description. – Shell moderately involute, with a thick, subtriangular whorl section. Flanks relatively flat or slightly convex, converging towards the narrowly rounded, slightly fastigate venter; ventral shoulders rounded; relatively abrupt umbilical shoulders and shallow umbilical wall. Ornamentation consisting of prorsiradiate ribs fading on upper flanks; growth striae slightly projected on the venter.

Measurements. – See Appendix.

Discussion. – This species differs from *Tropigastrites lahontanus* by having thicker whorls and more involute coiling.

Occurrence. – Fossil Hill (Humboldt Range): FHB 34 (2); *Rotelliformis* Zone (Late Anisian).

Superfamily Arcestaceae Mojsisovics, 1875

Family Arcestidae Mojsisovics, 1875

Genus *Proarcestes* Mojsisovics, 1893

Proarcestes cf. *P. bramantei* (Mojsisovics, 1869)

Figs. 51–53; Pl. 30: 6–9

1869 *Arcestes bramantei* – Mojsisovics, p. 575; Pl. 16: 1; Pl. 19: 4.

1882 *Arcestes bramantei* – Mojsisovics, p. 161; Pl. 46: 3–6.

1968 *Proarcestes* cf. *P. bramantei* (Mojsisovics) – Silberling & Tozer, p. 37.

1992b *Proarcestes* cf. *P. bramantei* (Mojsisovics) – Bucher, p. 442; Pl. 11: 15–20; Text-fig. 28.

Description. – Very involute, depressed, globose shell with an oval whorl section, thickest whorl width near the umbilicus, and high, slightly overhanging umbilical wall. Inner whorls more depressed than outer whorls. Shell smooth, with highly prorsiradiate, slightly sinuous striae. Phragmocone usually with three, irregularly spaced, internal varices per whorl. Length of body chamber exceeds one whorl. Suture line ammonitic, with an elongated ventral saddle and a unifid first lateral saddle.

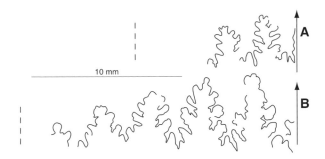

Fig. 51. Suture lines (×4) of *Proarcestes* cf. *P. bramantei* (Mojsisovics, 1869). A: PIMUZ 25359, Loc. HB 713, Oliver Gulch (Augusta Mountains); *Cordeyi* Subzone, *Weitschati* Zone, Late Anisian. B: PIMUZ 25354, Loc. HB 584, Oliver Gulch (Augusta Mountains); *Cordeyi* Subzone, *Weitschati* Zone, Late Anisian.

Measurements. – See Figures 52 and 53. Although highly variable, H, W, and U have a normal distribution and all display allometric growth. Whorl height is apparently quite variable, but the number of specimens available is insufficient to attribute this variation to intraspecific variation or to an evolutionary trend towards greater whorl height.

Discussion. – *Proarcestes* cf. *P. bramantei* differs from the stratigraphically higher *Proarcestes* cf. *P. balfouri* by having a parabolic ventral outline rather than a semicircular section. *Proarcestes* cf. *P. bramantei* differs from *Proarcestes gabbi* by having a more depressed whorl section. Moreover, as noted by Bucher (1992b), *Proarcestes* cf. *P. bramantei* differs from these two species by its suture line (unifid first lateral saddle).

Figure 53 compares our specimens, herein assigned to *Proarcestes* cf. *P. bramantei*, to those of Silberling & Nichols (1982). Obviously, it is apparent that *P. balfouri* can be distinguished by its more circular whorl section, whereas *Proarcestes* cf. *P. bramantei* and *P. gabbi* have a similar oval shape. However, on average, *Proarcestes* cf. *P. bramantei* is more depressed. As already noted by Bucher (1992b), the latter two species have somewhat different suture lines and they are also separated stratigraphically by two subzones (*Nevadanus* and *Meeki* subzones of the *Meeki* Zone). Note that none of our specimens can be attributed to *P. balfouri*, even those collected from the Fossil Hill section. Finally, it is noteworthy that the two specimens of *Proarcestes* cf. *P. bramantei* from the *Shoshonensis* Zone figured by Bucher (1992b) fall within the range of the younger specimens herein assigned to *Proarcestes* cf. *P. bramantei*.

Occurrence. – Oliver Gulch (Augusta Mountains): HB 584 (13), HB 585 (4), HB 592 (1), HB 593 (12), HB 596 (1), HB 597 (7), HB 598 (1), HB 705 (2), HB 707 (1), HB 710 (1), HB 713 (6), HB 2040 (1), HB 2052 (2); *Weitschati*

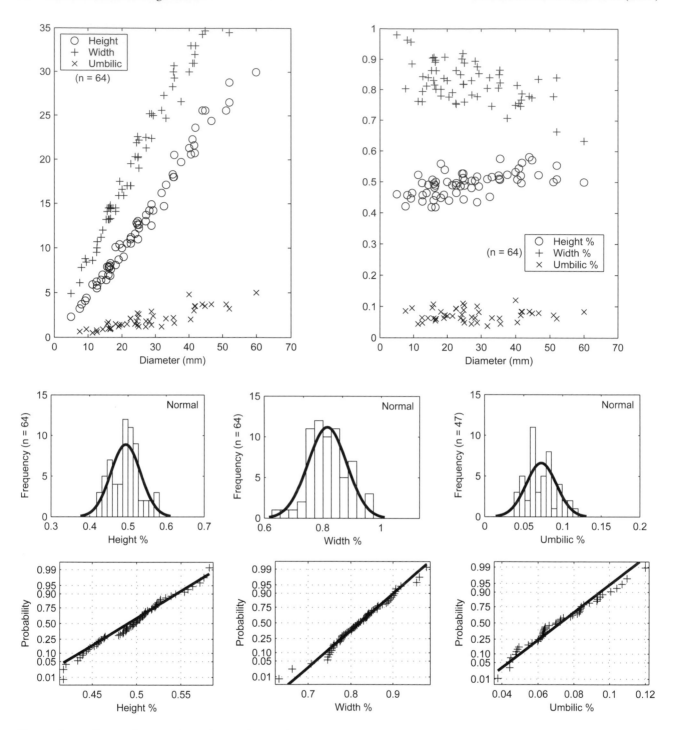

Fig. 52.   Scatter diagram of H, W, and U, and scatter diagram, histograms, and probability plots of H/D, W/D, and U/D for *Proarcestes* cf. *P. bramantei* (Mojsisovics, 1869) (Augusta Mountains; *Weitschati* and *Mimetus* Zones, Late Anisian).

Zone (Late Anisian). Ferguson Canyon (Augusta Mountains): HB 751 (1?), HB 2027 (1), HB 2034 (1); *Weitschati* Zone (Late Anisian). Muller Canyon (Augusta Mountains): HB 717 (2), HB 735 (1), HB 737 (1); *Mimetus* and *Rotelliformis* Zones (Late Anisian). Rieber Gulch (Augusta Mountains): HB 2004 (1), HB 2007 (4),

HB2014 (2), HB 2016 (2); *Weitschati* Zone (Late Anisian). McCoy Mine (New Pass Range): HB 2062 (10); *Weitschati* Zone (Late Anisian). Fossil Hill (Humboldt Range): FHB 6 (3), FHB 8 (15), FHB 9 (17), FHB 12 (1), FHB 15A (7), FHB 16 (2), FHB 28 (1), FHB 30 (2); *Rotelliformis* Zone (Late Anisian).

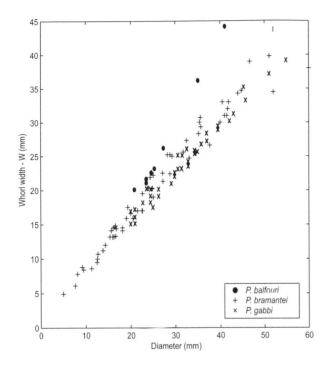

Fig. 53. Scatter diagram of W for the genus *Proarcestes*, split by species (*P.* cf. *P. bramantei*, *P. balfouri*, and *P. gabbi*). Data for *P. balfouri* and *P. gabbi* from Silberling & Nichols (1982).

Order Phylloceratida Arkell, 1950

Superfamily Phyllocerataceae Zittel, 1884

Family Ussuritidae Hyatt, 1900

Genus *Ussurites* Hyatt, 1900

Ussurites cf. *U. arthaberi* (Welter, 1915)

Figs. 54, 55; Pl. 4: 8–11

1870 *Ammonites billingsianus* – Gabb, p. 8; Pl. 5: 3.
1905 *Monophyllites billingsianus* – Hyatt & Smith, p. 94; Pl. 24: 3–4.
1914 *Monophyllites billingsianus* – Smith, p. 48; Pl. 5: 3–4.
1915 *Monophyllites arthaberi* – Welter, p. 115; Pl. 89: 1a–c.
1934 *Ussurites billingsianus* – Spath, p. 286; fig. 100i.
1982 *Ussurites* cf. *arthaberi* – Silberling & Nichols, p. 60; Pl. 32: 3–4; Text-fig. 46.

Description. – Depressed to high oval, evolute shell with a rounded venter, convex flanks (greatest width low on the flanks), and a high, convex umbilical wall. Ornamentation smooth or composed of numerous, straight or slightly convex or sinuous, rectiradiate radial lirae. Whorl section initially very depressed on innermost whorls,

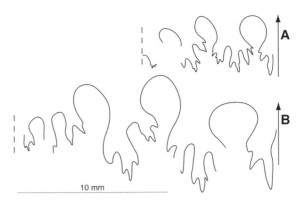

Fig. 54. Suture lines (× 4) of *Ussurites* cf. *U. arthaberi* (Welter, 1915). A: PIMUZ 25462, Loc. HB 2010, Rieber Gulch (Augusta Mountains); *Cordeyi* Subzone, *Weitschati* Zone, Late Anisian; reversed. B: PIMUZ 25463, Loc. HB 730, Muller Canyon (Augusta Mountains); *Cordeyi* Subzone, *Weitschati* Zone, Late Anisian.

becoming progressively higher and compressed throughout ontogeny, leading to a high oval outer whorl section. Innermost whorls also bear megastriae and conspicuous lateral, parabolic nodes. Suture line typically has distinct, phylloid saddles, with the first lateral saddle E/L lacking subdivisions on the ventral side.

Measurements. – See Figure 55. Note the progressive compression of the whorl section throughout ontogeny.

Discussion. – Although the presence of lateral parabolic lines on the depressed innermost whorls of our specimens is apparently unknown, the shape and the suture line are very similar to those figured by Silberling & Nichols (1982) and Welter (1915). However, the marginal parabolic lines on the outer whorls of Welter's holotype are not present on our specimens.

Occurrence. – Oliver Gulch (Augusta Mountains): HB 584 (2), HB 585 (2), HB 713 (1); *Cordeyi* Subzone, *Weitschati* Zone (Late Anisian). Ferguson Canyon (Augusta Mountains): HB 738 (3), HB 2034 (1); *Cordeyi* Subzone, *Weitschati* Zone (Late Anisian). Muller Canyon (Augusta Mountains): HB 730 (1); *Cordeyi* Subzone, *Weitschati* Zone (Late Anisian). Rieber Gulch (Augusta Mountains): HB 2006 (1), HB 2007 (3), HB 2010 (1), HB 2014 (1); *Cordeyi* Subzone, *Weitschati* Zone (Late Anisian).

Acknowledgements

Field work in northwestern Nevada (USA) was supported by the Universität Stiftung of the University of Zürich. The authors are greatly indebted to Jim Jenks (Salt Lake City) for his assistance in the field, for the loan of comparative material, and for improving the English text. Hans Rieber (Zürich) is thanked for his help in the field and for fruitful discussions about Triassic ammonoid taxonomy. Technical support for photography and preparation was provided by Heinz Lanz, Markus Hebeisen,

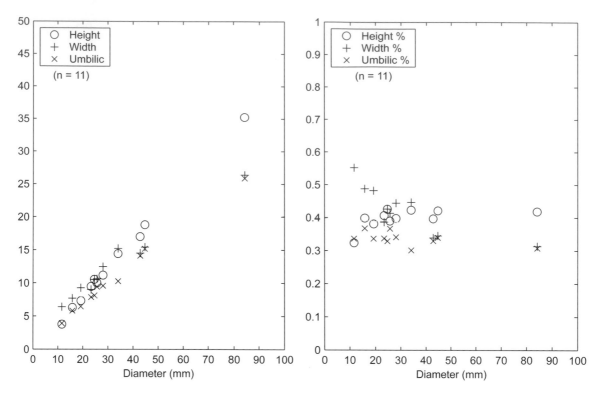

Fig. 55. Scatter diagram of H, W, and U, and of H/D, W/D, and U/D for *Ussurites* cf. *U. arthaberi* (Welter, 1915) (Augusta Mountains; Late Anisian).

Julia Huber, and Leonie Pauli. This paper is a contribution to the Swiss National Science Foundation project 200020-105090/1 (H. Bucher).

References

Arkell, W.J. 1950: A classification of the Jurassic ammonites. *Journal of Paleontology 24*, 354–364.

Arthaber, G. v. 1896: Die Cephalopodenfauna der Reiflinger Kalke. *Beitrage zur Palaeontologie und Geologischen Oesterreich-Ungarns und des Orients 10* (1–2, 4), 1–112, 192–242.

Arthaber, G. v. 1911: Die Trias von Albanien. *Beitrage zur Palaeontologie und Geologischen Oesterreich-Ungarns und des Orients 24*, 169–277.

Arthaber, G. v. 1912: Ueber die Horizontierung der Fossilfunde am Monte Cucco (italienische Carnia) und über die systematische Stellung von *Cuccoceras* Dien. *Jahrbuch der geologischen Bundesanstalt 62*, 333–358.

Balini, M. 1992a: *Lardaroceras* gen. n., a new Late Anisian ammonoid genus from the Prezzo Limestone (Southern Alps). *Rivista Italiana di Paleontologia e Stratigrafia 98*, 3–28.

Balini, M. 1992b: New genera of Anisian ammonoids from the Prezzo Limestone (Southern Alps). *Atti Ticinensi di Scienze della Terra 35*, 179–198.

Beyrich, E. 1867: *Uber einige Cephalopoden aus dem Muschelkalk der Alpen und über verwandte Arten*. Dümmler, Berlin, 45 pp.

Böckh, J. 1874: Die geologischen Verhaltnisse des sudlichen Theiles des Bakony, 11. Hungary, *Földtani Intézet Evkönyve 3*, 1–180.

Bucher, H. 1988: A new Middle Anisian (Middle Triassic) ammonoid zone from northwestern Nevada (USA). *Eclogae geologicae Helvetiae 81*, 723–762.

Bucher, H. 1989: Lower Anisian ammonoids from the northern Humboldt Range (northwestern Nevada, USA) and their bearing upon the Lower/Middle Triassic boundary. *Eclogae geologicae Helvetiae 82*, 945–1002.

Bucher, H. 1992a: Ammonoids of the *Hyatti* Zone and the Anisian transgression in the Triassic Star Peak Group, northwestern Nevada, USA. *Palaeontographica, Abt. A 223*, 137–166.

Bucher, H. 1992b: Ammonoids of the *Shoshonensis* Zone (Middle Anisian, Middle Triassic) from northwestern Nevada (USA). *Jahrbuch der geologischen Bundesanstalt 135*, 425–465.

Bucher, H. 1994: New ammonoids from the *Taylori* Zone (Middle Anisian, Middle Triassic) from northwestern Nevada (USA). *Mémoire de Géologie (Lausanne) 22*, 1–8.

Dagys, A. & Weitschat, W. 1993: Extensive intraspecific variation in a Triassic ammonoid from Siberia. *Lethaia 26*, 113–121.

Diener, C. 1916: Einige Bemerkungen zur Nomenklatur der Trias-cephalopoden. *Zentralblat für mineralogie, geologie und paläontologie*, Stuttgart, 97–105.

Gabb, W.M. 1864: Description of the Triassic fossils of California and the adjacent territories. *California Geological Survey, Paleontology 1*, 17–35.

Gabb, W.M. 1870: Description of some secondary fossils from the Pacific states. *American Journal of Conchology 5*, 5–18.

Guex, J. 1991: *Biochronological correlations*. Springer-Verlag, Berlin, 255 pp.

Hauer, F. v. 1896: Beitrage zur Kenntniss der Cephalopoden aus der Trias von Bosnien, II. Nautilen und Ammoniten mit ceratitischen Loben aus dem Muschelkalk von Haliluci bei Sarajevo. *Akademie der Wissenschaften in Wien, Mathematich-naturwissenschaftlichen Klasse, Denkschriften 63*, 237–276.

Hyatt, A. 1877: Paleontology (Trias). *U.S. Geological Exploration of the 40th Parallel 4*, 124.

Hyatt, A. 1884: Genera of fossil cephalopods. *Proceedings of the Boston Society of Natural History 22*, 253–338.

Hyatt, A. 1900. Cephalopoda. *In* Zittel, K.A. (ed.): *Textbook of Palaeontology*, C.R. Eastman, London, 502–592.

Hyatt, A. & Smith, J.P. 1905: The Triassic cephalopod genera of America. *U.S. Geological Survey Professional Paper 40*, 394 pp.

Kiparisova, L.D. 1958: Ptychitaceae. *In* Luppov, N.P. & Drushchits, V.V. (eds.): *Fundamentals of Paleontology, Mollusca – Cephalopoda II, Ammonoidea (Ceratitida, Ammonitida, etc.)*, 50–52 [in Russian].

McLearn, F.H. 1969: Middle Triassic (Anisian) ammonoids from northeastern British Columbia and Ellesmere Island. *Geological Survey of Canada Paper 66-54*, 3 pp.

Meek, F.B. 1877: Paleontology (Trias). *U.S. Geological Exploration of the 40th Parallel 4*, 1–197.

Mojsisovics, E. v. 1869: Ueber die Gliederung der oberen Triasbildungen der oestlichen Alpen. *Jahrbuch der geologischen Reichsanstalt Wien 19*, 91–150.

Mojsisovics, E. v. 1873: Das Gebirge um Hallstatt, theil I, Die Molluscanfaunen der Zlambach- und Hallstätter-Schichten. *Abhandlungen der Kaiserlich-Königlichen Geologischen Reichsanstalt 6 (1)*, 1–82.

Mojsisovics, E. v. 1875: Das Gebirge um Hallstatt, theil I, Die Molluscanfaunen der Zlambach- und Hallstätter-Schichten. *Abhandlungen der Kaiserlich-Königlichen Geologischen Reichsanstalt 6 (1)*, 83–174.

Mojsisovics, E. v. 1879: Vorlaufige kurze übersicht der Ammoniten-Gattungen der mediterranen und juvavischen Trias. *Verhandlungen der geologischen Reichsanstalt Wien, Jahrgang 1879*, 133–143.

Mojsisovics, E. v. 1882: Die Cephalopoden der mediterranen Triasprovinz. *Abhandlungen der Kaiserlich-Königlichen Geologischen Reichsanstalt 10*, 322 pp.

Mojsisovics, E. v. 1893: Das Gebirge um Hallstatt, Theil I, Die Cephalopoden der Hallstätter Kalke. *Abhandlungen der Kaiserlich-Königlichen Geologischen Reichsanstalt 6 (2)*, 835 pp.

Monnet, C. & Bucher, H. 2002: Cenomanian (early Late Cretaceous) ammonoid faunas of Western Europe. Part I: biochronology (Unitary Associations) and diachronism of datums. *Eclogae geologicae Helvetiae 95*, 57–73.

Nichols, K.M. & Silberling, N.J. 1977: Stratigraphy and depositional history of the Star Peak Group (Triassic), northwestern Nevada. *Geological Society of America Special Paper 178*, 73 pp.

Silberling, N.J. 1962: Stratigraphic distribution of Middle Triassic ammonites at Fossil Hill, Humboldt Range, Nevada. *Journal of Paleontology 36 (1)*, 153–160.

Silberling, N.J. & Nichols, K.M. 1982: Middle Triassic molluscan fossils of biostratigraphic significance from the Humboldt Range,

north-western Nevada. *U.S. Geological Survey Professional Paper 1207*, 77 pp.

Silberling, N.J. & Tozer, E.T. 1968: Biostratigraphic classification of the marine Triassic in North America. *Geological Society of America Special Paper 110*, 63 pp.

Silberling, N.J. & Wallace, R.E. 1969: Stratigraphy of the Star Peak Group (Triassic) and overlying lower Mesozoic rocks, Humboldt Range, Nevada. *U.S. Geological Survey Professional Paper 592*, 50 pp.

Smith, J.P. 1904: The comparative stratigraphy of the marine Trias of western America. *California Academy of Science Proceedings, ser. 3, 1*, 323–430.

Smith, J.P. 1914: The Middle Triassic marine invertebrate faunas of North America. *U.S. Geological Survey Professional Paper 83*, 254 pp.

Spath, L.F. 1934: *Catalogue of the fossil cephalopoda in the British Museum (National History). Part 4, The ammonoidea of the Trias.* London, 521 pp.

Spath, L.F. 1951: *Catalogue of the fossil Cephalopoda in the British Museum (Natural History). Part 5, The Ammonoids of the Trias (II).* London, 228 pp.

Tatzreiter, F. & Balini, M. 1993: The new genus *Schreyerites* and its type species *Ceratites abichi* Mojsisovics, 1882 (Ammonoidea, Anisian, Middle Triassic). *Atti Ticinensi di Scienze della Terra 36*, 1–10.

Tozer, E.T. 1971: Triassic time and ammonoids – Problems and proposals. *Canadian Journal of Earth Sciences 8 (8)*, 989–1031. Errata and addenda, p. 1611.

Tozer, E.T. 1981: Triassic Ammonoidea: classification, evolution and relationship with Permian and Jurassic forms. *In* House, M.R. & Senior, J.R. (eds.): The Ammonoidea. *Systematics Association Special Volume 18*, 65–100.

Tozer, E.T. 1994: Canadian Triassic ammonoid faunas. *Geological Survey of Canada, Bulletin 467*, 663 pp.

Waagen, W. 1895: Fossils from the Ceratite Formation. *Palaeontologia Indica 13 (2)*, 323 pp.

Welter, O.A. 1915: Die Ammoniten und Nautiliden der ladinischen und anisischen Trias von Timor. – *Paläontologie von Timor V*, Schweizentartschen Verlags buch handlung Stuttgart, 71–136.

Westermann, G.E.G. 1966: Covariation and taxonomy of the Jurassic ammonite Sanninia adicra (Waagen). Neues Jahrbuch für Geologie und Paläontologie Abhandlungen 124, 289–312.

Wyld, S.J. 2000: Triassic evolution of the arc and backarc of northwestern Nevada, and evidence for extensional tectonism. *Geological Society of America Special Paper 347*, 185–207.

Zittel, K.A. v. 1884: *Handbuch der Palaeontologie. Cephalopoda.* Druck und Verlag von R. Oldenbourg, Munich, 239–522.

Appendix

Species	Specimen	D (mm)	H/D (%)	W/D (%)	U/D (%)
Sageceras walteri Mojsisovics, 1882	PIMUZ 25425	33.1	57.7	19.3	4.4
	PIMUZ 25424	79.5	58.2	17.9	4.8
	PIMUZ 25426	85.1	58.4	20.1	6.9
Acrochordiceras carolinae Mojsisovics, 1882	PIMUZ 25129	21.6	47.2	44.0	25.2
	PIMUZ 25130	19.1	49.5	62.3	27.2
	PIMUZ 25127	41.6	52.4	39.4	16.6
	PIMUZ 25128	80.2	49.4	38.5	22.9
	HB 741-2	15.6	46.2	47.8	27.9
	PIMUZ 25126	27.9	49.5	45.3	22.2
Balatonites hexatuberculatus n. sp.	PIMUZ 25134, Holotype	78.0*	37.1*	31.3*	45.8*
	PIMUZ 25135	12.0	35.0	19.2	45.0
Dixieceras lawsoni (Smith, 1914)	Holotype (after Smith 1914)	74.0	47.3	31.1	22.3
Bulogites mojsvari Arthaber, 1896	PIMUZ 25204	66.8	49.6	25.9	20.5
	PIMUZ 25203	66.6	47.1	30.3	22.4
	USNM 452800	83.3	42.9	31.2*	26.9
Marcouxites spinifer (Smith, 1914)	Holotype (after Smith 1914)	60.0	43.3	36.7	30.8
	FHB 11 – 45455	66.7	45.9	40.8	22.0
Silberlingia praecursor n. sp.	PIMUZ 25449, Holotype	71.2	41.7	26.4	28.1
Brackites spinosus n. sp.	PIMUZ 25183, Holotype	34.4	43.3	36.9	33.7
	PIMUZ 25182, Paratype	46.6	41.6	35.0	33.5
Eutomoceras dunni Smith, 1904	PIMUZ 25261	33.3	50.4	32.4	19.8
	PIMUZ 25259	38.1	49.8	27.6	18.9
	PIMUZ 25258	44.5	48.5	22.0	18.9
	PIMUZ 25260	19.5	44.6	29.7	26.7
Tropigymnites sp. indet.	PIMUZ 25456	24.0	38.3	34.6	37.5
	PIMUZ 25457	23.3	39.5	31.8	36.9
Gymnites sp. indet.	PIMUZ 25262	37.6	33.8	19.9	41.0
	PIMUZ 25264	22.4	39.7	18.8	36.6
	PIMUZ 25263	20.0	38.0	18.0	39.0
Anagymnites sp. indet.	PIMUZ 25131	56.5	41.8	21.2	30.3
Discoptychites cf. *D. megalodiscus* Beyrich, 1867	HB 591-1	185.0	56.8	25.4	3.8
	PIMUZ 25218	97.0	58.8	32.0	4.1
Ptychites sp. indet.	HB 2062-9	22.8	50.9	59.6	18.4
	PIMUZ 25362	43.7	51.7	52.4	16.0
Oxylongobardites acutus n. sp.	PIMUZ 25352, Holotype	25.6	58.2*	15.6	–
	PIMUZ 25351	18.7	60.4	19.8	3.2
	HB 2036-1	22.2	64.0	22.1	3.2
Tropigastrites lahontanus Smith, 1914	PIMUZ 25452	33.2	26.2	28.6	53.0
	PIMUZ 25453	40.0	30.0	21.3	50.0
	FHB 17 – 45546	26.8	33.6	26.9	46.6
Tropigastrites louderbacki (Hyatt & Smith, 1905)	PIMUZ 25455	37.0	28.1	33.8	49.5

* Estimated values

Plates 1–31

Plate 1

1: Aerial view towards south of the northwestern slopes of the Augusta Mountains. Favret Canyon is at the centre of the view.

2: Northern slope of Ferguson Canyon, Augusta Mountains. MB 2–6 indicate the location of the marker beds in this area (see Fig. 4). MB 2 lies between the *Shoshonensis* Zone and the new *Weitschati* Zone. MB 3–4 are within the *Lawsoni* Subzone (*Mimetus* Zone). MB 6 is at the base of the *Vogdesi* Subzone (*Rotelliformis* Zone).

3: Weathered cross-section of bed HB 2032 (MB 3) in the Ferguson West section (Fig. 5), Augusta Mountains. Note the numerous sections of *Dixieceras lawsoni* (*Mimetus* Zone, Late Anisian), which also outline some soft sediment deformation.

4: Early diagenetic calcareous nodule (HB 735) with *Dixieceras lawsoni* (*Mimetus* Zone, Late Anisian) from Muller Canyon, Augusta Mountains.

5: View towards south of Rieber Gulch section, Augusta Mountains. The small cliff at the bottom of the section (white arrow) contains ammonites of the *Ransomei* Subzone (*Shoshonensis* Zone, Middle Anisian). The top of the hill (black arrow) is within the *Weitschati* Zone of the Late Anisian (sample HB 2006; see Fig. 6).

6: View of the locality HB 2016 (*Weitschati* Zone, Late Anisian) in Rieber Gulch, Augusta Mountains (see Fig. 6).

PLATE 1

FOSSILS AND STRATA 52 (2005) 61

Plate 2 (all figures natural size)

1a–b: ***Balatonites hexatuberculatus* n. sp.** PIMUZ 25134, holotype.
Loc. HB 739, Ferguson Canyon (Augusta Mountains). *Mojsvari* Subzone, *Shoshonensis* Zone, Middle Anisian.

2a-b: ***Balatonites hexatuberculatus*? n. sp.** PIMUZ 25135.
Loc. HB 739, Ferguson Canyon (Augusta Mountains). *Mojsvari* Subzone, *Shoshonensis* Zone, Middle Anisian.

3a–b: ***Platycuccoceras* sp. indet.** PIMUZ 25353.
Loc. HB 739, Ferguson Canyon (Augusta Mountains). *Mojsvari* Subzone, *Shoshonensis* Zone, Middle Anisian.

4a–c: ***Chiratites bituberculatus* n. gen. n. sp.** PIMUZ 25213, holotype.
Loc. HB 739, Ferguson Canyon (Augusta Mountains). *Mojsvari* Subzone, *Shoshonensis* Zone, Middle Anisian.

5a–b: ***Acrochordiceras carolinae* Mojsisovics, 1882.** PIMUZ 25127.
Loc. HB 739, Ferguson Canyon (Augusta Mountains). *Mojsvari* Subzone, *Shoshonensis* Zone, Middle Anisian.

6a–b: ***Acrochordiceras carolinae* Mojsisovics, 1882.** PIMUZ 25128.
Loc. HB 739, Ferguson Canyon (Augusta Mountains). *Mojsvari* Subzone, *Shoshonensis* Zone, Middle Anisian.

7a–c: ***Acrochordiceras carolinae* Mojsisovics, 1882.** PIMUZ 25129.
Loc. HB 739, Ferguson Canyon (Augusta Mountains). *Mojsvari* Subzone, *Shoshonensis* Zone, Middle Anisian.

8a–c: ***Acrochordiceras carolinae* Mojsisovics, 1882.** PIMUZ 25130.
Loc. HB 739, Ferguson Canyon (Augusta Mountains). *Mojsvari* Subzone, *Shoshonensis* Zone, Middle Anisian.

9a-c: ***Acrochordiceras carolinae* Mojsisovics, 1882.** PIMUZ 25126.
Loc. HB 741, Oliver Gulch (Augusta Mountains). *Mojsvari* Subzone, *Shoshonensis* Zone, Middle Anisian.

PLATE 2
FOSSILS AND STRATA 52 (2005)   63

Plate 3 (all figures natural size)

1a–b: ***Chiratites retrospinosus* n. gen. n. sp.** PIMUZ 25214, holotype. Loc. HB 739, Ferguson Canyon (Augusta Mountains). *Mojsvari* Subzone, *Shoshonensis* Zone, Middle Anisian.

2a–c: ***Chiratites retrospinosus* n. gen. n. sp.** PIMUZ 25215, paratype. Loc. HB 739, Ferguson Canyon (Augusta Mountains). *Mojsvari* Subzone, *Shoshonensis* Zone, Middle Anisian.

3a–c: ***Chiratites retrospinosus* n. gen. n. sp.** PIMUZ 25216, paratype. Loc. HB 739, Ferguson Canyon (Augusta Mountains). *Mojsvari* Subzone, *Shoshonensis* Zone, Middle Anisian.

4a-c: ***Chiratites retrospinosus* n. gen. n. sp.** PIMUZ 25217, paratype. Loc. HB 739, Ferguson Canyon (Augusta Mountains). *Mojsvari* Subzone, *Shoshonensis* Zone, Middle Anisian.

PLATE 3

FOSSILS AND STRATA 52 (2005) 65

1a 1b 2a 2b 2c 3a 3b 3c 4a 4b 4c

Plate 4 (all figures natural size)

1a–b: ***Bulogites mojsvari* (Arthaber, 1896).** PIMUZ 25203.
Loc. HB 739, Ferguson Canyon (Augusta Mountains). *Mojsvari* Subzone, *Shoshonensis* Zone, Middle Anisian.

2a–b: ***Bulogites mojsvari* (Arthaber, 1896).** USNM 452800.
Muller Canyon (Augusta Mountains). *Mojsvari* Subzone, *Shoshonensis* Zone, Middle Anisian. Specimen illustrated by Bucher (1992b; Text-fig. 17).

3a–b: ***Bulogites mojsvari* (Arthaber, 1896).** PIMUZ 25204.
Loc. HR ##A, McCoy Mine (New Pass Range). *Mojsvari* Subzone, *Shoshonensis* Zone, Middle Anisian.

4a–b: ***Bulogites mojsvari* (Arthaber, 1896).** PIMUZ 25205.
Loc. HB 739, Ferguson Canyon (Augusta Mountains). *Mojsvari* Subzone, *Shoshonensis* Zone, Middle Anisian.

5a–b: ***Gymnites* sp. indet.** PIMUZ 25262.
Loc. HB 739, Ferguson Canyon (Augusta Mountains). *Mojsvari* Subzone, *Shoshonensis* Zone, Middle Anisian.

6a–b: ***Gymnites* sp. indet.** PIMUZ 25263.
Loc. HB 739, Ferguson Canyon (Augusta Mountains). *Mojsvari* Subzone, *Shoshonensis* Zone, Middle Anisian.

7a–b: ***Gymnites* sp. indet.** PIMUZ 25264.
Loc. HB 739, Ferguson Canyon (Augusta Mountains). *Mojsvari* Subzone, *Shoshonensis* Zone, Middle Anisian.

8a–b: ***Ussurites* cf. *U. arthaberi* (Welter, 1915).** PIMUZ 25460.
Loc. HB 2007, Rieber Gulch (Augusta Mountains). *Cordeyi* Subzone, *Weitschati* Zone, Late Anisian.

9a–b: ***Ussurites* cf. *U. arthaberi* (Welter, 1915).** PIMUZ 25458.
Loc. HB 738, Ferguson Canyon (Augusta Mountains). *Cordeyi* Subzone, *Weitschati* Zone, Late Anisian.

10a–b: ***Ussurites* cf. *U. arthaberi* (Welter, 1915).** PIMUZ 25461.
Loc. HB 584, Oliver Gulch (Augusta Mountains). *Cordeyi* Subzone, *Weitschati* Zone, Late Anisian. Arrow indicates the location of a parabolic line.

11a–c: ***Ussurites* cf. *U. arthaberi* (Welter, 1915).** PIMUZ 25459.
Loc. HB 738, Ferguson Canyon (Augusta Mountains). *Cordeyi* Subzone, *Weitschati* Zone, Late Anisian.

PLATE 4

FOSSILS AND STRATA 52 (2005) 67

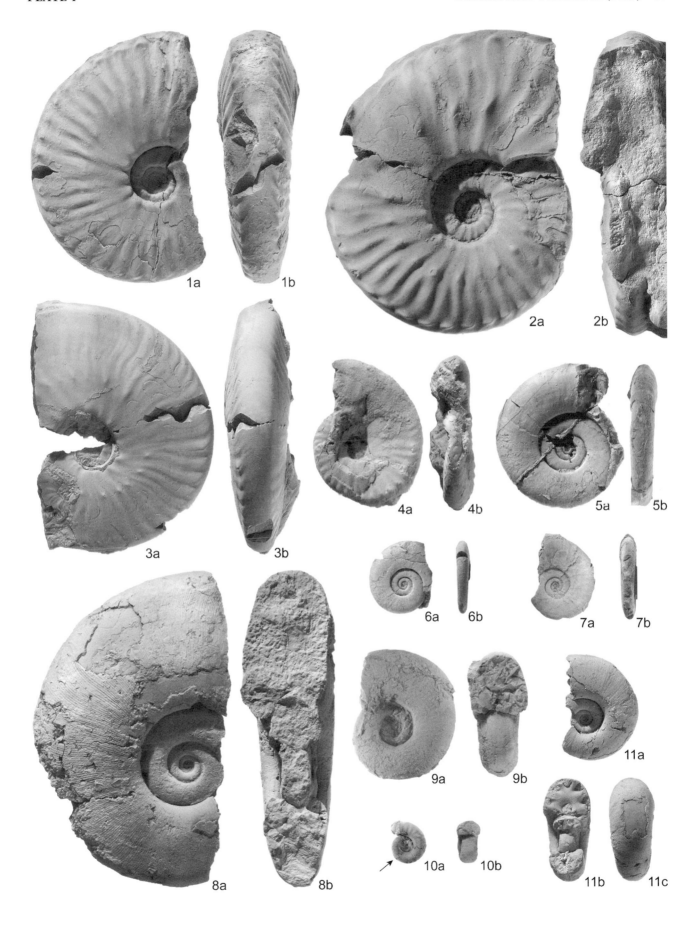

Plate 5 (all figures natural size)

1a–c: ***Billingsites cordeyi* n. gen. n. sp.** PIMUZ 25136, holotype.
Loc. HB 590, Oliver Gulch (Augusta Mountains). *Cordeyi* Subzone, *Weitschati* Zone, Late Anisian.

2a–c: ***Billingsites cordeyi* n. gen. n. sp.** PIMUZ 25137, paratype.
Loc. HB 584, Oliver Gulch (Augusta Mountains). *Cordeyi* Subzone, *Weitschati* Zone, Late Anisian.

3a–c: ***Billingsites cordeyi* n. gen. n. sp.** PIMUZ 25138, paratype.
Loc. HB 584, Rieber Gulch (Augusta Mountains). *Cordeyi* Subzone, *Weitschati* Zone, Late Anisian.

4a–c: ***Billingsites cordeyi* n. gen. n. sp.** PIMUZ 25139, paratype.
Loc. HB 584, Oliver Gulch (Augusta Mountains). *Cordeyi* Subzone, *Weitschati* Zone, Late Anisian.

5a–b: ***Billingsites cordeyi* n. gen. n. sp.** PIMUZ 25140, paratype.
Loc. HB 584, Oliver Gulch (Augusta Mountains). *Cordeyi* Subzone, *Weitschati* Zone, Late Anisian.

6a–b: ***Billingsites cordeyi* n. gen. n. sp.** PIMUZ 25141.
Loc. HB 585, Oliver Gulch (Augusta Mountains). *Cordeyi* Subzone, *Weitschati* Zone, Late Anisian.

PLATE 5

FOSSILS AND STRATA 52 (2005) 69

1a 1b 1c

2a 2b 2c

3a 3b 3c

4a 4b 4c

5a 5b

6a 6b

Plate 6 (all figures natural size)

1a–c: ***Billingsites cordeyi* n. gen. n. sp.** PIMUZ 25142, paratype. Loc. HB 584, Oliver Gulch (Augusta Mountains). *Cordeyi* Subzone, *Weitschati* Zone, Late Anisian.

2a–c: ***Billingsites cordeyi* n. gen. n. sp.** PIMUZ 25143, paratype. Loc. HB 590, Oliver Gulch (Augusta Mountains). *Cordeyi* Subzone, *Weitschati* Zone, Late Anisian.

3a–c: ***Billingsites cordeyi* n. gen. n. sp.** PIMUZ 25144, paratype. Loc. HB 584, Oliver Gulch (Augusta Mountains). *Cordeyi* Subzone, *Weitschati* Zone, Late Anisian.

4a–c: ***Billingsites cordeyi* n. gen. n. sp.** PIMUZ 25145, paratype. Loc. HB 584, Oliver Gulch (Augusta Mountains). *Cordeyi* Subzone, *Weitschati* Zone, Late Anisian.

5a–c: ***Billingsites cordeyi* n. gen. n. sp.** PIMUZ 25146, paratype. Loc. HB 590, Oliver Gulch (Augusta Mountains). *Cordeyi* Subzone, *Weitschati* Zone, Late Anisian.

6a–c: ***Billingsites cordeyi* n. gen. n. sp.** PIMUZ 25147, paratype. Loc. HB 584, Oliver Gulch (Augusta Mountains). *Cordeyi* Subzone, *Weitschati* Zone, Late Anisian.

7a–c: ***Billingsites cordeyi* n. gen. n. sp.** PIMUZ 25148, paratype. Loc. HB 584, Oliver Gulch (Augusta Mountains). *Cordeyi* Subzone, *Weitschati* Zone, Late Anisian.

8a–c: ***Billingsites cordeyi* n. gen. n. sp.** PIMUZ 25149, paratype. Loc. HB 590, Oliver Gulch (Augusta Mountains). *Cordeyi* Subzone, *Weitschati* Zone, Late Anisian.

9a–b: ***Billingsites cordeyi* n. gen. n. sp.** PIMUZ 25150, paratype. Loc. HB 584, Ferguson Canyon (Augusta Mountains). *Cordeyi* Subzone, *Weitschati* Zone, Late Anisian.

10a–b: ***Billingsites cordeyi* n. gen. n. sp.** PIMUZ 25151, paratype. Loc. HB 584, Oliver Gulch (Augusta Mountains). *Cordeyi* Subzone, *Weitschati* Zone, Late Anisian.

11a–b: ***Billingsites cordeyi* n. gen. n. sp.** PIMUZ 25152, paratype. Loc. HB 590, Oliver Gulch (Augusta Mountains). *Cordeyi* Subzone, *Weitschati* Zone, Late Anisian.

12a–b: ***Billingsites cordeyi* n. gen. n. sp.** PIMUZ 25158. Loc. HB 2021, Ferguson Canyon (Augusta Mountains). *Cordeyi* Subzone, *Weitschati* Zone, Late Anisian. Injured specimen developing a pathological ventral ribbing.

13a–c: ***Billingsites cordeyi* n. gen. n. sp.** PIMUZ 25153, paratype. Loc. HB 590, Oliver Gulch (Augusta Mountains). *Cordeyi* Subzone, *Weitschati* Zone, Late Anisian.

PLATE 6

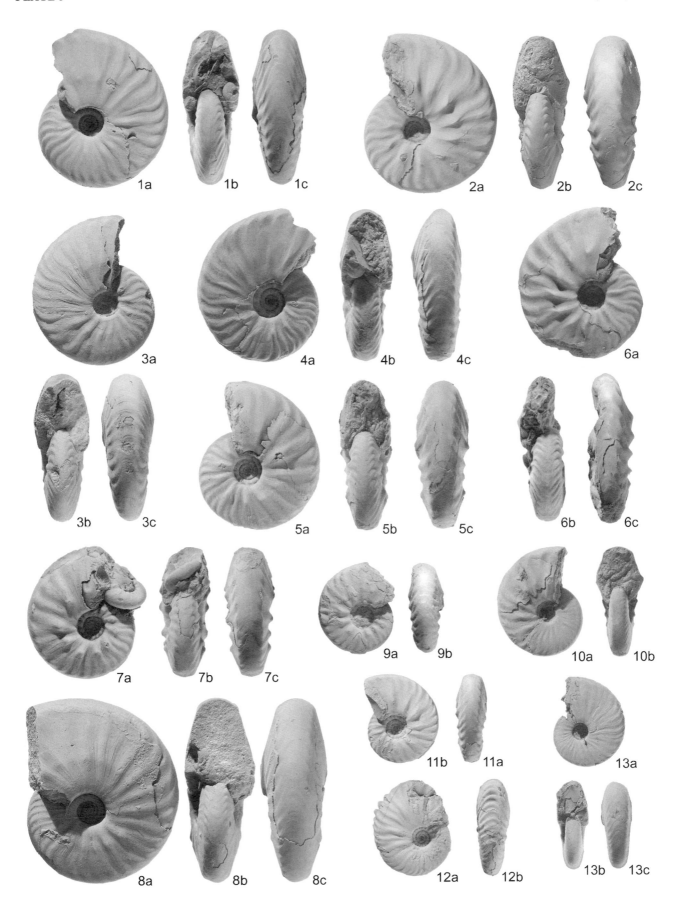

1a   1b   1c   2a   2b   2c

3a   4a   4b   4c   6a

3b   3c   5a   5b   5c   6b   6c

7a   7b   7c   9a   9b   10a   10b

11b   11a   13a

8a   8b   8c   12a   12b   13b   13c

Plate 7 (all figures natural size)

1a–b: ***Billingsites cordeyi*** **n. gen. n. sp.** PIMUZ 25154.
Loc. HB 2021, Ferguson Canyon (Augusta Mountains). *Cordeyi* Subzone, *Weitschati* Zone, Late Anisian.

2a–c: ***Billingsites cordeyi*** **n. gen. n. sp.** PIMUZ 25155, paratype.
Loc. HB 590, Oliver Gulch (Augusta Mountains). *Cordeyi* Subzone, *Weitschati* Zone, Late Anisian.

3a–c: ***Billingsites cordeyi*** **n. gen. n. sp.** PIMUZ 25156.
Loc. HB 2033, Ferguson West (Augusta Mountains). *Cordeyi* Subzone, *Weitschati* Zone, Late Anisian.

4a–b: ***Billingsites cordeyi*** **n. gen. n. sp.** PIMUZ 25157.
Loc. HB 740, Ferguson Canyon (Augusta Mountains). *Cordeyi* Subzone, *Weitschati* Zone, Late Anisian.

5a–b: ***Billingsites escargueli*** **n. gen. n. sp.** PIMUZ 25181, paratype.
Loc. HB 738, Ferguson Canyon (Augusta Mountains). *Cordeyi* Subzone, *Weitschati* Zone, Late Anisian.

6a–c: ***Billingsites cordeyi*** **n. gen. n. sp.** PIMUZ 25159.
Loc. HB 2033, Ferguson West (Augusta Mountains). *Cordeyi* Subzone, *Weitschati* Zone, Late Anisian.

PLATE 7

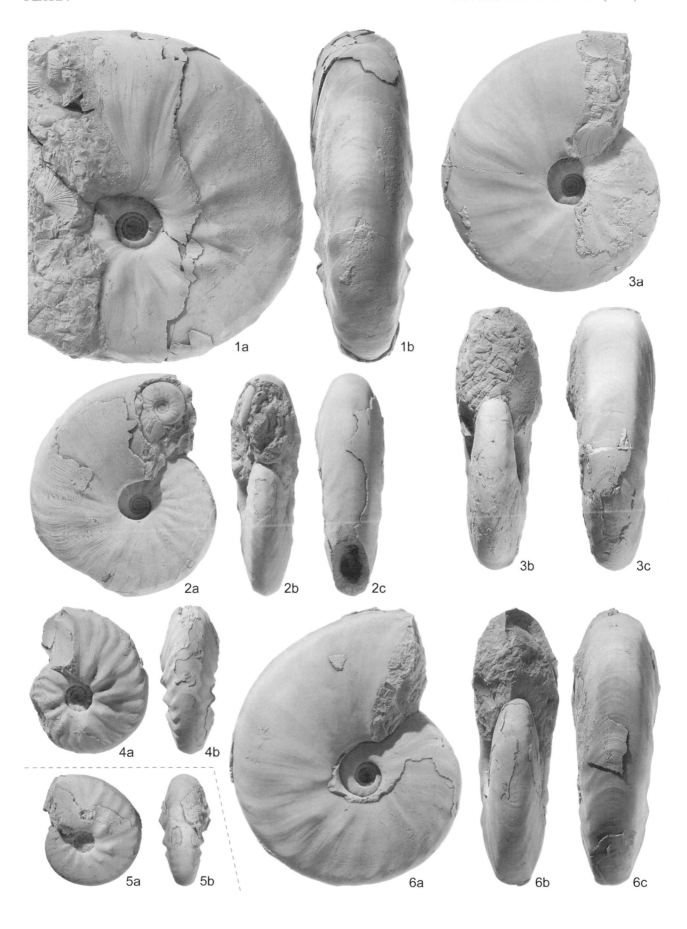

Plate 8 (all figures natural size)

1a–b: ***Billingsites escargueli* n. gen. n. sp.** PIMUZ 25165, paratype. Loc. HB 738, Ferguson Canyon (Augusta Mountains). *Cordeyi* Subzone, *Weitschati* Zone, Late Anisian.

2a–c: ***Billingsites escargueli* n. gen. n. sp.** PIMUZ 25166, holotype. Loc. HB 738, Ferguson Canyon (Augusta Mountains). *Cordeyi* Subzone, *Weitschati* Zone, Late Anisian.

3a–c: ***Billingsites escargueli* n. gen. n. sp.** PIMUZ 25167, paratype. Loc. HB 738, Ferguson Canyon (Augusta Mountains). *Cordeyi* Subzone, *Weitschati* Zone, Late Anisian.

4a–c: ***Billingsites escargueli* n. gen. n. sp.** PIMUZ 25164. Loc. HB 2015, Rieber Gulch (Augusta Mountains). *Cordeyi* Subzone, *Weitschati* Zone, Late Anisian.

5a–c: ***Billingsites escargueli* n. gen. n. sp.** PIMUZ 25168, paratype. Loc. HB 738, Ferguson Canyon (Augusta Mountains). *Cordeyi* Subzone, *Weitschati* Zone, Late Anisian.

6a–c: ***Billingsites escargueli* n. gen. n. sp.** PIMUZ 25169, paratype. Loc. HB 738, Ferguson Canyon (Augusta Mountains). *Cordeyi* Subzone, *Weitschati* Zone, Late Anisian.

7a–b: ***Billingsites escargueli* n. gen. n. sp.** PIMUZ 25170, paratype. Loc. HB 738, Ferguson Canyon (Augusta Mountains). *Cordeyi* Subzone, *Weitschati* Zone, Late Anisian.

8a–b: ***Billingsites escargueli* n. gen. n. sp.** PIMUZ 25171, paratype. Loc. HB 738, Ferguson Canyon (Augusta Mountains). *Cordeyi* Subzone, *Weitschati* Zone, Late Anisian.

9a–c: ***Billingsites escargueli* n. gen. n. sp.** PIMUZ 25172, paratype. Loc. HB 738, Ferguson Canyon (Augusta Mountains). *Cordeyi* Subzone, *Weitschati* Zone, Late Anisian.

10a–b: ***Billingsites escargueli* n. gen. n. sp.** PIMUZ 25173, paratype. Loc. HB 738, Ferguson Canyon (Augusta Mountains). *Cordeyi* Subzone, *Weitschati* Zone, Late Anisian.

11a–b: ***Billingsites escargueli* n. gen. n. sp.** PIMUZ 25174, paratype. Loc. HB 738, Ferguson Canyon (Augusta Mountains). *Cordeyi* Subzone, *Weitschati* Zone, Late Anisian.

12a–b: ***Billingsites escargueli* n. gen. n. sp.** PIMUZ 25175, paratype. Loc. HB 738, Ferguson Canyon (Augusta Mountains). *Cordeyi* Subzone, *Weitschati* Zone, Late Anisian.

13a–b: ***Billingsites escargueli* n. gen. n. sp.** PIMUZ 25176, paratype. Loc. HB 738, Ferguson Canyon (Augusta Mountains). *Cordeyi* Subzone, *Weitschati* Zone, Late Anisian.

PLATE 8

FOSSILS AND STRATA 52 (2005)   75

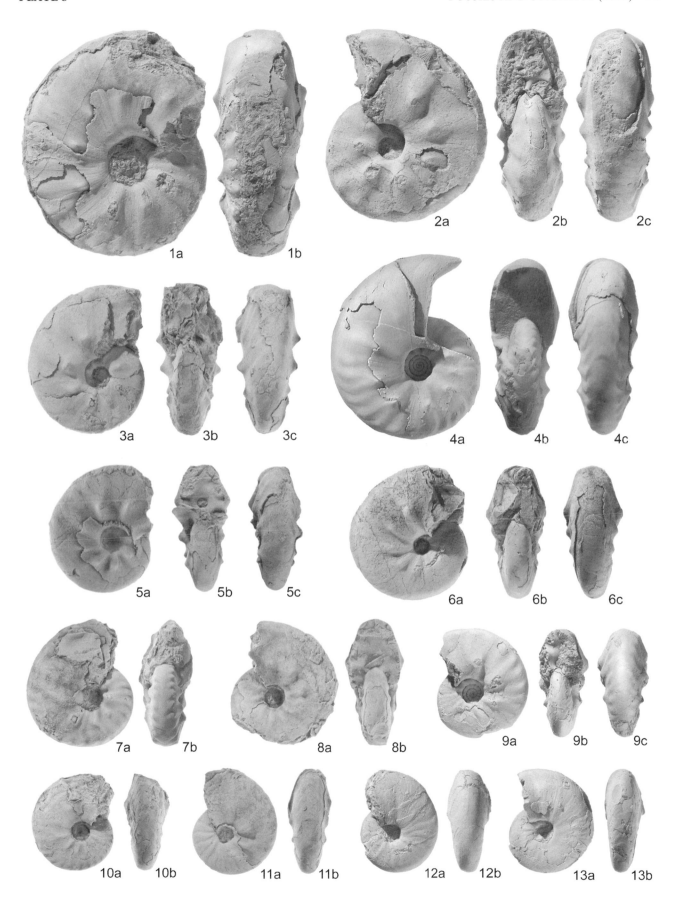

Plate 9 (all figures natural size)

1a–b: ***Billingsites escargueli* n. gen. n. sp.** PIMUZ 25177, paratype. Loc. HB 738, Ferguson Canyon (Augusta Mountains). *Cordeyi* Subzone, *Weitschati* Zone, Late Anisian.

2a–b: ***Billingsites escargueli* n. gen. n. sp.** PIMUZ 25178. Loc. HB 740, Ferguson Canyon (Augusta Mountains). *Cordeyi* Subzone, *Weitschati* Zone, Late Anisian.

3a–b: ***Billingsites escargueli* n. gen. n. sp.** PIMUZ 25179, paratype. Loc. HB 738, Ferguson Canyon (Augusta Mountains). *Cordeyi* Subzone, *Weitschati* Zone, Late Anisian.

4a–c: ***Marcouxites spinifer* (Smith, 1914).** PIMUZ 25349. Loc. FHB 11, Fossil Hill (Humboldt Range). *Spinifer* Subzone, *Mimetus* Zone, Late Anisian.

5a: ***Marcouxites spinifer* (Smith, 1914).** PIMUZ 25350. Loc. FHB 11, Fossil Hill (Humboldt Range). *Spinifer* Subzone, *Mimetus* Zone, Late Anisian.

6a–b: ***Silberlingia clarkei* (Smith, 1914).** PIMUZ 25427. Loc. HB 736, Muller Canyon (Augusta Mountains). Late Anisian.

7a–b: ***Ceccaceras stecki* n. gen. n. sp.** PIMUZ 25206, paratype. Loc. FHB 18, Fossil Hill (Humboldt Range). *Vogdesi* Subzone, *Rotelliformis* Zone, Late Anisian.

8a–c: ***Ceccaceras stecki* n. gen. n. sp.** PIMUZ 25208, paratype. Loc. FHB 18, Fossil Hill (Humboldt Range). *Vogdesi* Subzone, *Rotelliformis* Zone, Late Anisian.

9a–b: ***Ceccaceras stecki* n. gen. n. sp.** PIMUZ 25207, holotype. Loc. FHB 18, Fossil Hill (Humboldt Range). *Vogdesi* Subzone, *Rotelliformis* Zone, Late Anisian.

10a–c: ***Ceccaceras stecki* n. gen. n. sp.** PIMUZ 25209, paratype. Loc. FHB 18, Fossil Hill (Humboldt Range). *Vogdesi* Subzone, *Rotelliformis* Zone, Late Anisian.

11a–b: ***Ceccaceras stecki* n. gen. n. sp.** PIMUZ 25210, paratype. Loc. FHB 18, Fossil Hill (Humboldt Range). *Vogdesi* Subzone, *Rotelliformis* Zone, Late Anisian.

12a–b: ***Ceccaceras stecki* n. gen. n. sp.** PIMUZ 25211, paratype. Loc. FHB 18, Fossil Hill (Humboldt Range). *Vogdesi* Subzone, *Rotelliformis* Zone, Late Anisian.

13a–b: ***Ceccaceras stecki* n. gen. n. sp.** PIMUZ 25212, paratype. Loc. FHB 18, Fossil Hill (Humboldt Range). *Vogdesi* Subzone, *Rotelliformis* Zone, Late Anisian.

PLATE 9

FOSSILS AND STRATA 52 (2005)   77

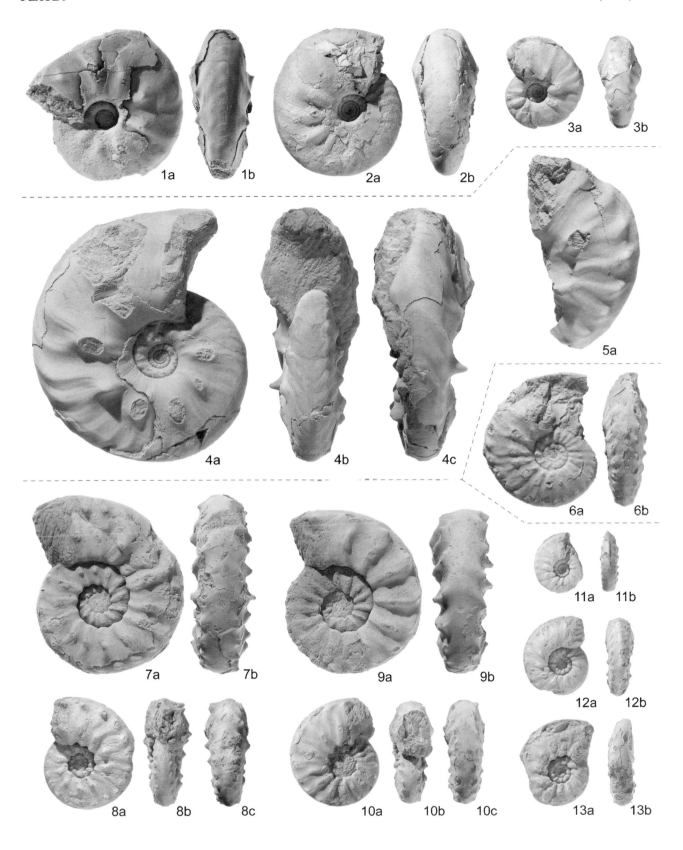

Plate 10 (all figures natural size)

1a–b: ***Dixieceras lawsoni* (Smith, 1914).** PIMUZ 25219.
Loc. FHB 9, Fossil Hill (Humboldt Range). *Lawsoni* Subzone,
Mimetus Zone, Late Anisian.

2a–b: ***Dixieceras lawsoni* (Smith, 1914).** PIMUZ 25220.
Loc. FHB 8, Fossil Hill (Humboldt Range). *Lawsoni* Subzone,
Mimetus Zone, Late Anisian.

3a–b: ***Dixieceras lawsoni* (Smith, 1914).** PIMUZ 25221.
Loc. FHB 8, Fossil Hill (Humboldt Range). *Lawsoni* Subzone,
Mimetus Zone, Late Anisian.

4a–c: ***Dixieceras lawsoni* (Smith, 1914).** PIMUZ 25222.
Loc. FHB 9, Fossil Hill (Humboldt Range). *Lawsoni* Subzone,
Mimetus Zone, Late Anisian.

5a–c: ***Dixieceras lawsoni* (Smith, 1914).** PIMUZ 25224.
Loc. FHB 8, Fossil Hill (Humboldt Range). *Lawsoni* Subzone,
Mimetus Zone, Late Anisian.

6a–c: ***Dixieceras lawsoni* (Smith, 1914).** PIMUZ 25223.
Loc. FHB 8, Fossil Hill (Humboldt Range). *Lawsoni* Subzone,
Mimetus Zone, Late Anisian.

7a–c: ***Dixieceras lawsoni* (Smith, 1914).** PIMUZ 25226.
Loc. FHB 9, Fossil Hill (Humboldt Range). *Lawsoni* Subzone,
Mimetus Zone, Late Anisian.

8a–c: ***Dixieceras lawsoni* (Smith, 1914).** PIMUZ 25230.
Loc. FHB 9, Fossil Hill (Humboldt Range). *Lawsoni* Subzone,
Mimetus Zone, Late Anisian.

9a–b: ***Dixieceras lawsoni* (Smith, 1914).** PIMUZ 25225.
Loc. FHB 9, Fossil Hill (Humboldt Range). *Lawsoni* Subzone,
Mimetus Zone, Late Anisian.

10a–b: ***Dixieceras lawsoni* (Smith, 1914).** PIMUZ 25227.
Loc. FHB 9, Fossil Hill (Humboldt Range). *Lawsoni* Subzone,
Mimetus Zone, Late Anisian.

11a–b: ***Dixieceras lawsoni* (Smith, 1914).** PIMUZ 25228.
Loc. FHB 9, Fossil Hill (Humboldt Range). *Lawsoni* Subzone,
Mimetus Zone, Late Anisian.

12a–b: ***Dixieceras lawsoni* (Smith, 1914).** PIMUZ 25229.
Loc. FHB 9, Fossil Hill (Humboldt Range). *Lawsoni* Subzone,
Mimetus Zone, Late Anisian.

PLATE 10

FOSSILS AND STRATA 52 (2005) 79

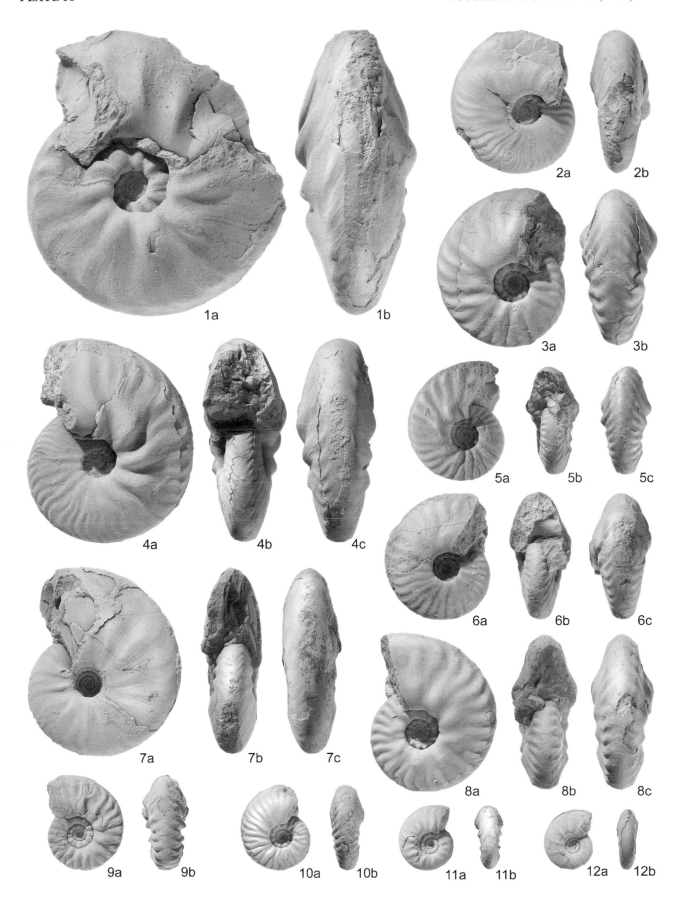

Plate 11 (all figures natural size)

1a–c: ***Dixieceras lawsoni* (Smith, 1914).** PIMUZ 25231.
Loc. HB 735, Muller Canyon (Augusta Mountains). *Lawsoni* Subzone, *Mimetus* Zone, Late Anisian.

2a–c: ***Dixieceras lawsoni* (Smith, 1914).** PIMUZ 25232.
Loc. HB 717, Muller Canyon (Augusta Mountains). *Lawsoni* Subzone, *Mimetus* Zone, Late Anisian.

3a–b: ***Dixieceras lawsoni* (Smith, 1914).** PIMUZ 25233.
Loc. HB 735, Muller Canyon (Augusta Mountains). *Lawsoni* Subzone, *Mimetus* Zone, Late Anisian.

4a–c: ***Dixieceras lawsoni* (Smith, 1914).** PIMUZ 25240.
Loc. HB 711, Oliver Gulch (Augusta Mountains). *Lawsoni* Subzone, *Mimetus* Zone, Late Anisian.

5a–c: ***Dixieceras lawsoni* (Smith, 1914).** PIMUZ 25234.
Loc. HB 735, Muller Canyon (Augusta Mountains). *Lawsoni* Subzone, *Mimetus* Zone, Late Anisian.

6a–b: ***Dixieceras lawsoni* (Smith, 1914).** PIMUZ 25241.
Loc. HB 717, Muller Canyon (Augusta Mountains). *Lawsoni* Subzone, *Mimetus* Zone, Late Anisian.

7a–b: ***Dixieceras lawsoni* (Smith, 1914).** PIMUZ 25235.
Loc. HB 735, Muller Canyon (Augusta Mountains). *Lawsoni* Subzone, *Mimetus* Zone, Late Anisian.

8a–b: ***Dixieceras lawsoni* (Smith, 1914).** PIMUZ 25236.
Loc. HB 735, Muller Canyon (Augusta Mountains). *Lawsoni* Subzone, *Mimetus* Zone, Late Anisian.

9a–b: ***Dixieceras lawsoni* (Smith, 1914).** PIMUZ 25242.
Loc. HB 710, Oliver Gulch (Augusta Mountains). *Lawsoni* Subzone, *Mimetus* Zone, Late Anisian.

10a–b: ***Dixieceras lawsoni* (Smith, 1914).** PIMUZ 25237.
Loc. HB 735, Muller Canyon (Augusta Mountains). *Lawsoni* Subzone, *Mimetus* Zone, Late Anisian.

11a–c: ***Dixieceras lawsoni* (Smith, 1914).** PIMUZ 25238.
Loc. HB 735, Muller Canyon (Augusta Mountains). *Lawsoni* Subzone, *Mimetus* Zone, Late Anisian.

12a–c: ***Dixieceras lawsoni* (Smith, 1914).** PIMUZ 25239.
Loc. HB 735, Muller Canyon (Augusta Mountains). *Lawsoni* Subzone, *Mimetus* Zone, Late Anisian.

PLATE 11

FOSSILS AND STRATA 52 (2005)   81

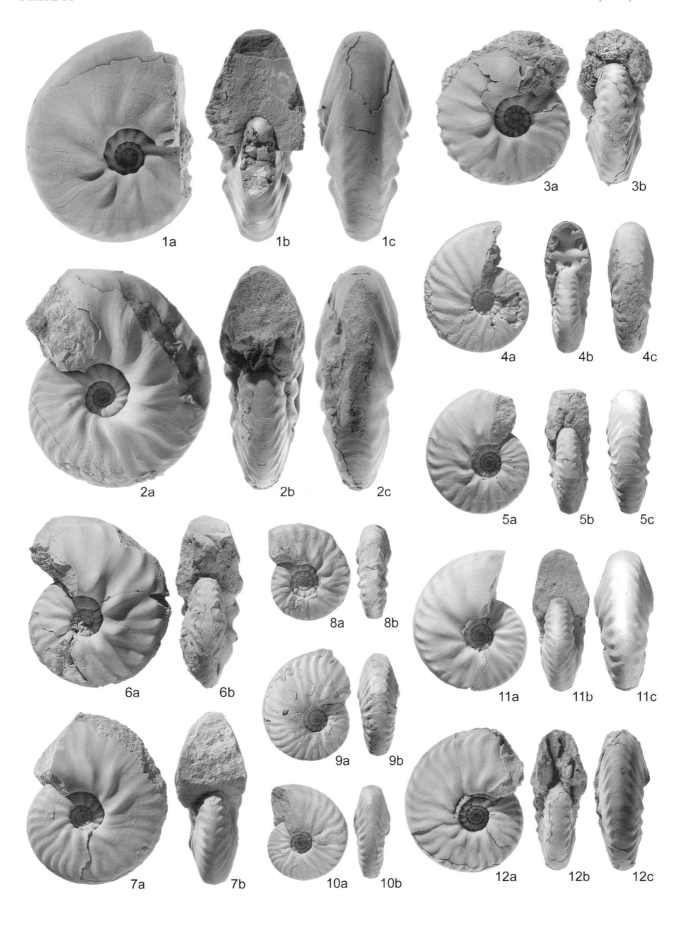

Plate 12 (all figures natural size)

1a–b: ***Dixieceras lawsoni* (Smith, 1914).** PIMUZ 25243.
   Loc. HB 597, Oliver Gulch (Augusta Mountains). *Lawsoni* Subzone, *Mimetus* Zone, Late Anisian.
2a–c: ***Dixieceras lawsoni* (Smith, 1914).** PIMUZ 25244.
   Loc. HB 597, Oliver Gulch (Augusta Mountains). *Lawsoni* Subzone, *Mimetus* Zone, Late Anisian.
3a–c: ***Dixieceras lawsoni* (Smith, 1914).** PIMUZ 25245.
   Loc. HB 597, Oliver Gulch (Augusta Mountains). *Lawsoni* Subzone, *Mimetus* Zone, Late Anisian.
4a–b: ***Dixieceras lawsoni* (Smith, 1914).** PIMUZ 25246.
   Loc. HB 597, Oliver Gulch (Augusta Mountains). *Lawsoni* Subzone, *Mimetus* Zone, Late Anisian.
5a–b: ***Dixieceras lawsoni* (Smith, 1914).** PIMUZ 25247.
   Loc. HB 597, Oliver Gulch (Augusta Mountains). *Lawsoni* Subzone, *Mimetus* Zone, Late Anisian.
6a–c: ***Dixieceras lawsoni* (Smith, 1914).** PIMUZ 25248.
   Loc. HB 597, Oliver Gulch (Augusta Mountains). *Lawsoni* Subzone, *Mimetus* Zone, Late Anisian.
7a–c: ***Dixieceras lawsoni* (Smith, 1914).** PIMUZ 25249.
   Loc. HB 597, Oliver Gulch (Augusta Mountains). *Lawsoni* Subzone, *Mimetus* Zone, Late Anisian.
8a–c: ***Dixieceras lawsoni* (Smith, 1914).** PIMUZ 25250.
   Loc. HB 597, Oliver Gulch (Augusta Mountains). *Lawsoni* Subzone, *Mimetus* Zone, Late Anisian.
9a–c: ***Dixieceras lawsoni* (Smith, 1914).** PIMUZ 25251.
   Loc. HB 597, Oliver Gulch (Augusta Mountains). *Lawsoni* Subzone, *Mimetus* Zone, Late Anisian.

PLATE 12

FOSSILS AND STRATA 52 (2005)   83

Plate 13 (all figures natural size)

1a–c: ***Gymnotoceras rotelliformis* Meek, 1877.** PIMUZ 25276.
Loc. FHB 22, Fossil Hill (Humboldt Range). *Vogdesi* Subzone, *Rotelliformis* Zone, Late Anisian.

2a–b: ***Gymnotoceras rotelliformis* Meek, 1877.** PIMUZ 25277.
Loc. FHB 22, Fossil Hill (Humboldt Range). *Vogdesi* Subzone, *Rotelliformis* Zone, Late Anisian.

3a–c: ***Gymnotoceras rotelliformis* Meek, 1877.** PIMUZ 25278.
Loc. FHB 22, Fossil Hill (Humboldt Range). *Vogdesi* Subzone, *Rotelliformis* Zone, Late Anisian.

4a–b: ***Gymnotoceras rotelliformis* Meek, 1877.** PIMUZ 25279.
Loc. FHB 22, Fossil Hill (Humboldt Range). *Vogdesi* Subzone, *Rotelliformis* Zone, Late Anisian.

5a–b: ***Gymnotoceras rotelliformis* Meek, 1877.** PIMUZ 25280.
Loc. FHB 18, Fossil Hill (Humboldt Range). *Vogdesi* Subzone, *Rotelliformis* Zone, Late Anisian.

6a–b: ***Gymnotoceras rotelliformis* Meek, 1877.** PIMUZ 25281.
Loc. FHB 18, Fossil Hill (Humboldt Range). *Vogdesi* Subzone, *Rotelliformis* Zone, Late Anisian.

7a–b: ***Gymnotoceras rotelliformis* Meek, 1877.** PIMUZ 25284.
Loc. FHB 15a, Fossil Hill (Humboldt Range). *Vogdesi* Subzone, *Rotelliformis* Zone, Late Anisian.

8a–b: ***Gymnotoceras rotelliformis* Meek, 1877.** PIMUZ 25285.
Loc. FHB 15a, Fossil Hill (Humboldt Range). *Vogdesi* Subzone, *Rotelliformis* Zone, Late Anisian.

9a–b: ***Gymnotoceras rotelliformis* Meek, 1877.** PIMUZ 25286.
Loc. FHB 15a, Fossil Hill (Humboldt Range). *Vogdesi* Subzone, *Rotelliformis* Zone, Late Anisian.

10a–c: ***Gymnotoceras rotelliformis* Meek, 1877.** PIMUZ 25287.
Loc. FHB 12, Fossil Hill (Humboldt Range). *Vogdesi* Subzone, *Rotelliformis* Zone, Late Anisian.

11a–b: ***Gymnotoceras rotelliformis* Meek, 1877.** PIMUZ 25282.
Loc. FHB 18, Fossil Hill (Humboldt Range). *Vogdesi* Subzone, *Rotelliformis* Zone, Late Anisian.

12a–c: ***Gymnotoceras rotelliformis* Meek, 1877.** PIMUZ 25289.
Loc. HB 742, Oliver Gulch (Augusta Mountains). *Vogdesi* Subzone, *Rotelliformis* Zone, Late Anisian.

13a–b: ***Gymnotoceras rotelliformis* Meek, 1877.** PIMUZ 25283.
Loc. FHB 17, Fossil Hill (Humboldt Range). *Vogdesi* Subzone, *Rotelliformis* Zone, Late Anisian.

14a–b: ***Gymnotoceras rotelliformis* Meek, 1877.** PIMUZ 25288.
Loc. FHB 15a, Fossil Hill (Humboldt Range). *Vogdesi* Subzone, *Rotelliformis* Zone, Late Anisian.

15a–c: ***Gymnotoceras rotelliformis* Meek, 1877.** PIMUZ 25290.
Loc. HB 712, Oliver Gulch (Augusta Mountains). *Vogdesi* Subzone, *Rotelliformis* Zone, Late Anisian.

PLATE 13

FOSSILS AND STRATA 52 (2005) 85

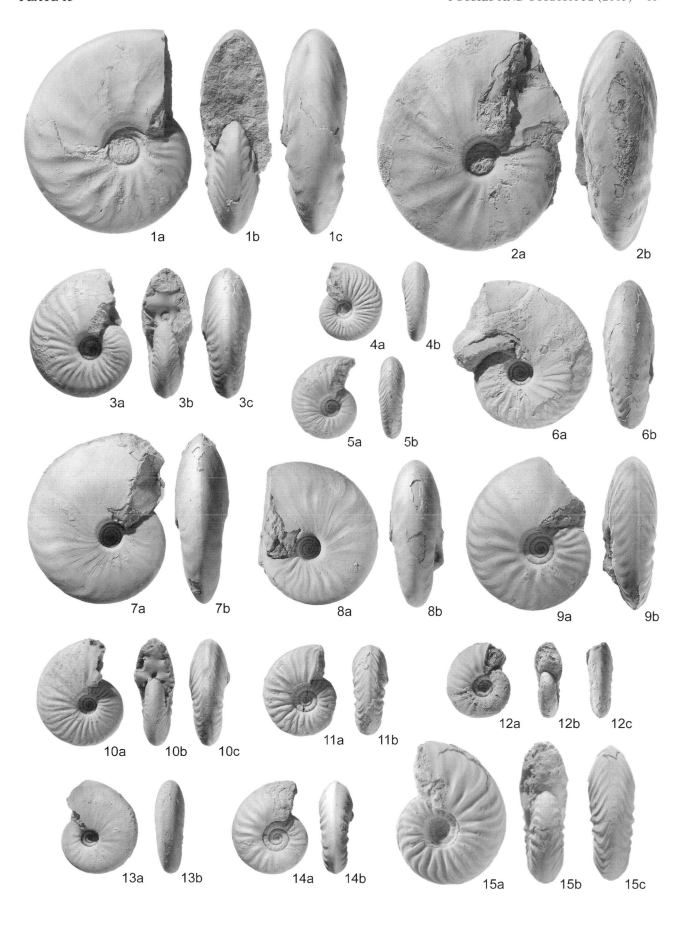

Plate 14 (all figures natural size)

1a–c: ***Gymnotoceras weitschati* n. sp.** PIMUZ 25292, holotype. Loc. HB 2006, Rieber Gulch (Augusta Mountains). *Cordeyi* Subzone, *Weitschati* Zone, Late Anisian.

2a–b: ***Gymnotoceras weitschati* n. sp.** PIMUZ 25293, paratype. Loc. HB 2006, Rieber Gulch (Augusta Mountains). *Cordeyi* Subzone, *Weitschati* Zone, Late Anisian.

3a–b: ***Gymnotoceras weitschati* n. sp.** PIMUZ 25294, paratype. Loc. HB 2006, Rieber Gulch (Augusta Mountains). *Cordeyi* Subzone, *Weitschati* Zone, Late Anisian.

4a–b: ***Gymnotoceras weitschati* n. sp.** PIMUZ 25295, paratype. Loc. HB 2006, Rieber Gulch (Augusta Mountains). *Cordeyi* Subzone, *Weitschati* Zone, Late Anisian.

5a–b: ***Gymnotoceras weitschati* n. sp.** PIMUZ 25296, paratype. Loc. HB 2006, Rieber Gulch (Augusta Mountains). *Cordeyi* Subzone, *Weitschati* Zone, Late Anisian.

6a–c: ***Gymnotoceras weitschati* n. sp.** PIMUZ 25297, paratype. Loc. HB 2006, Rieber Gulch (Augusta Mountains). *Cordeyi* Subzone, *Weitschati* Zone, Late Anisian.

7a–b: ***Gymnotoceras weitschati* n. sp.** PIMUZ 25298, paratype. Loc. HB 2006, Rieber Gulch (Augusta Mountains). *Cordeyi* Subzone, *Weitschati* Zone, Late Anisian.

8a–b: ***Gymnotoceras weitschati* n. sp.** PIMUZ 25299, paratype. Loc. HB 2006, Rieber Gulch (Augusta Mountains). *Cordeyi* Subzone, *Weitschati* Zone, Late Anisian.

9a–b: ***Gymnotoceras weitschati* n. sp.** PIMUZ 25300, paratype. Loc. HB 2006, Rieber Gulch (Augusta Mountains). *Cordeyi* Subzone, *Weitschati* Zone, Late Anisian.

10a–b: ***Gymnotoceras weitschati* n. sp.** PIMUZ 25301, paratype. Loc. HB 2006, Rieber Gulch (Augusta Mountains). *Cordeyi* Subzone, *Weitschati* Zone, Late Anisian.

11a–b: ***Gymnotoceras weitschati* n. sp.** PIMUZ 25302, paratype. Loc. HB 2006, Rieber Gulch (Augusta Mountains). *Cordeyi* Subzone, *Weitschati* Zone, Late Anisian.

12a–b: ***Gymnotoceras weitschati* n. sp.** PIMUZ 25303, paratype. Loc. HB 2006, Rieber Gulch (Augusta Mountains). *Cordeyi* Subzone, *Weitschati* Zone, Late Anisian.

13a–b: ***Gymnotoceras weitschati* n. sp.** PIMUZ 25304, paratype. Loc. HB 2006, Rieber Gulch (Augusta Mountains). *Cordeyi* Subzone, *Weitschati* Zone, Late Anisian.

PLATE 14

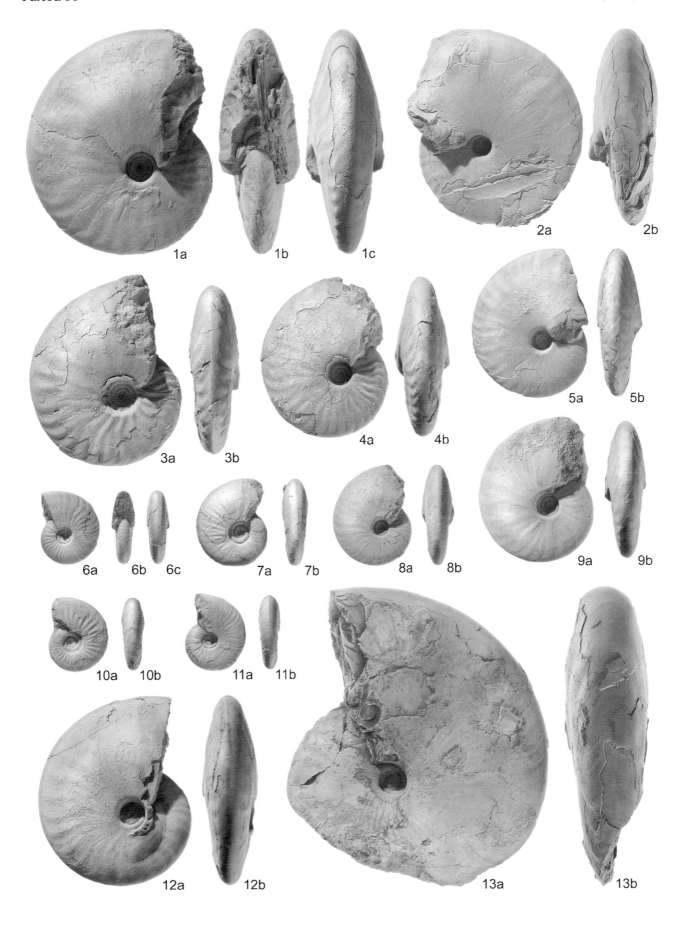

Plate 15 (all figures natural size)

1a–b: ***Gymnotoceras weitschati* n. sp.** PIMUZ 25305.
Loc. HB 2004, Rieber Gulch (Augusta Mountains). *Cordeyi* Subzone, *Weitschati* Zone, Late Anisian.

2a–c: ***Gymnotoceras weitschati* n. sp.** PIMUZ 25308.
Loc. HB 584, Oliver Gulch (Augusta Mountains). *Cordeyi* Subzone, *Weitschati* Zone, Late Anisian. Arrow indicates a parabolic line.

3a–c: ***Gymnotoceras weitschati* n. sp.** PIMUZ 25306.
Loc. HB 2004, Rieber Gulch (Augusta Mountains). *Cordeyi* Subzone, *Weitschati* Zone, Late Anisian.

4a–c: ***Gymnotoceras weitschati* n. sp.** PIMUZ 25309.
Loc. HB 713, Oliver Gulch (Augusta Mountains). *Cordeyi* Subzone, *Weitschati* Zone, Late Anisian.

5a–b: ***Gymnotoceras weitschati* n. sp.** PIMUZ 25307.
Loc. HB 2004, Rieber Gulch (Augusta Mountains). *Cordeyi* Subzone, *Weitschati* Zone, Late Anisian.

6a–c: ***Gymnotoceras weitschati* n. sp.** PIMUZ 25310.
Loc. HB 713, Oliver Gulch (Augusta Mountains). *Cordeyi* Subzone, *Weitschati* Zone, Late Anisian.

PLATE 15

FOSSILS AND STRATA 52 (2005)   89

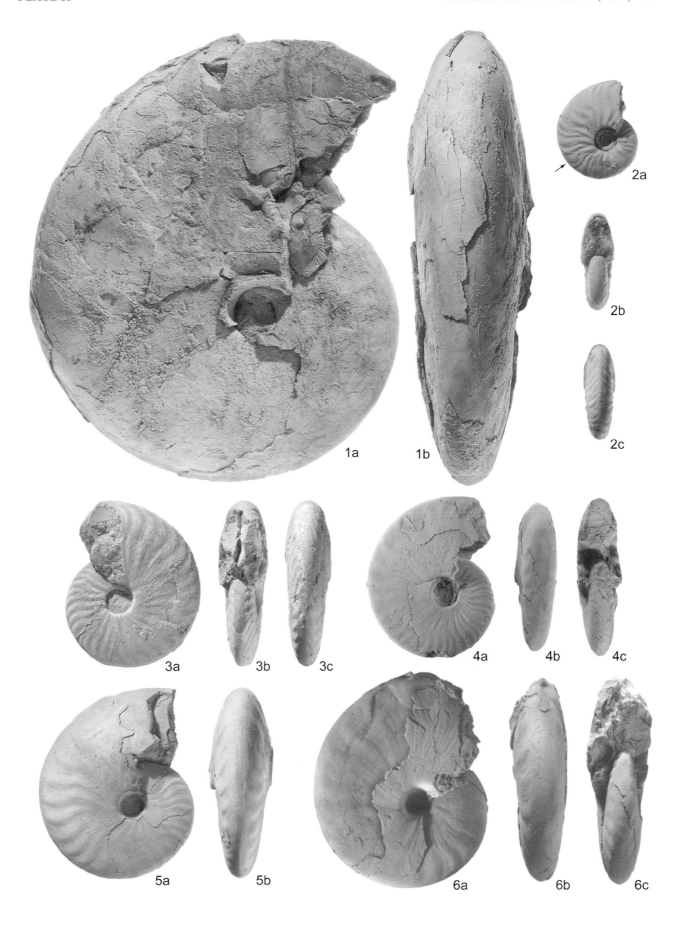

1a   1b   2a   2b   2c

3a   3b   3c   4a   4b   4c

5a   5b   6a   6b   6c

Plate 16 (all figures natural size)

1a–b: ***Gymnotoceras weitschati* n. sp.** PIMUZ 25314.
Loc. HB 596, Oliver Gulch (Augusta Mountains). *Transiformis*
Subzone, *Weitschati* Zone, Late Anisian.

2a–b: ***Gymnotoceras weitschati* n. sp.** PIMUZ 25315.
Loc. HB 2034, Ferguson West (Augusta Mountains). *Cordeyi*
Subzone, *Weitschati* Zone, Late Anisian.

3a–b: ***Gymnotoceras weitschati* n. sp.** PIMUZ 25313.
Loc. HB 596, Oliver Gulch (Augusta Mountains). *Transiformis*
Subzone, *Weitschati* Zone, Late Anisian.

4a–b: ***Gymnotoceras weitschati* n. sp.** PIMUZ 25312.
Loc. HB 2030, Ferguson West (Augusta Mountains). *Cordeyi*
Subzone, *Weitschati* Zone, Late Anisian.

5a–b: ***Gymnotoceras weitschati* n. sp.** PIMUZ 25316.
Loc. HB 2034, Ferguson West (Augusta Mountains). *Cordeyi*
Subzone, *Weitschati* Zone, Late Anisian.

6a–b: ***Gymnotoceras weitschati* n. sp.** PIMUZ 25317.
Loc. HB 2034, Ferguson West (Augusta Mountains). *Cordeyi*
Subzone, *Weitschati* Zone, Late Anisian.

7a–c: ***Gymnotoceras weitschati* n. sp.** PIMUZ 25318.
Loc. HB 2034, Ferguson West (Augusta Mountains). *Cordeyi*
Subzone, *Weitschati* Zone, Late Anisian.

8a–b: ***Gymnotoceras weitschati* n. sp.** PIMUZ 25319.
Loc. HB 2034, Ferguson West (Augusta Mountains). *Cordeyi*
Subzone, *Weitschati* Zone, Late Anisian.

9a–b: ***Gymnotoceras weitschati* n. sp.** PIMUZ 25311, paratype.
Loc. HB 2006, Rieber Gulch (Augusta Mountains). *Cordeyi*
Subzone, *Weitschati* Zone, Late Anisian.

10a–b: ***Sageceras walteri* Mojsisovics, 1882.** PIMUZ 25423.
Loc. FHB 8, Fossil Hill (Humboldt Range). *Lawsoni* Subzone,
Mimetus Zone, Late Anisian.

PLATE 16

FOSSILS AND STRATA 52 (2005)   91

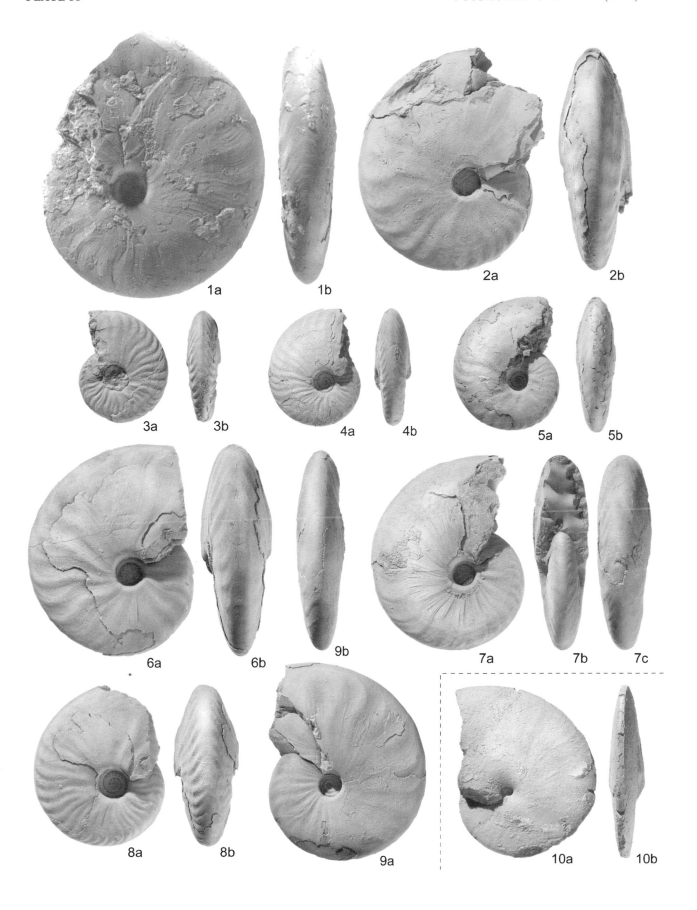

1a   1b   2a   2b

3a   3b   4a   4b   5a   5b

6a   6b   9b   7a   7b   7c

8a   8b   9a   10a   10b

Plate 17 (all figures natural size)

1a–c: ***Gymnotoceras mimetus* n. sp.** PIMUZ 25267, holotype.
Loc. FHB 9, Fossil Hill (Humboldt Range). *Lawsoni* Subzone, *Mimetus* Zone, Late Anisian.

2a–c: ***Gymnotoceras mimetus* n. sp.** PIMUZ 25268, paratype.
Loc. FHB 9, Fossil Hill (Humboldt Range). *Lawsoni* Subzone, *Mimetus* Zone, Late Anisian.

3a–c: ***Gymnotoceras mimetus* n. sp.** PIMUZ 25269, paratype.
Loc. FHB 9, Fossil Hill (Humboldt Range). *Lawsoni* Subzone, *Mimetus* Zone, Late Anisian.

4a–b: ***Gymnotoceras mimetus* n. sp.** PIMUZ 25270.
Loc. HB 735, Muller Canyon (Augusta Mountains). *Lawsoni* Subzone, *Mimetus* Zone, Late Anisian.

5a–b: ***Gymnotoceras mimetus* n. sp.** PIMUZ 25271.
Loc. HB 735, Muller Canyon (Augusta Mountains). *Lawsoni* Subzone, *Mimetus* Zone, Late Anisian.

6a–b: ***Gymnotoceras mimetus* n. sp.** PIMUZ 25275, paratype.
Loc. FHB 9, Fossil Hill (Humboldt Range). *Lawsoni* Subzone, *Mimetus* Zone, Late Anisian.

7a–b: ***Gymnotoceras mimetus* n. sp.** PIMUZ 25272.
Loc. HB 735, Muller Canyon (Augusta Mountains). *Lawsoni* Subzone, *Mimetus* Zone, Late Anisian.

8a–b: ***Gymnotoceras mimetus* n. sp.** PIMUZ 25273.
Loc. HB 735, Muller Canyon (Augusta Mountains). *Lawsoni* Subzone, *Mimetus* Zone, Late Anisian.

9a–c: ***Gymnotoceras mimetus* n. sp.** PIMUZ 25274.
Loc. HB 735, Muller Canyon (Augusta Mountains). *Lawsoni* Subzone, *Mimetus* Zone, Late Anisian.

10a–b: ***Anagymnites* sp. indet.** PIMUZ 25131.
Loc. FHB 8, Fossil Hill (Humboldt Range). *Lawsoni* Subzone, *Mimetus* Zone, Late Anisian.

11a–b: ***Anagymnites* sp. indet.** PIMUZ 25132.
Loc. HB 596, Oliver Gulch (Augusta Mountains). *Transiformis* Subzone, *Weitschati* Zone, Late Anisian.

12a–b: ***Anagymnites* sp. indet.** PIMUZ 25133.
Loc. FHB 11, Fossil Hill (Humboldt Range). *Spinifer* Subzone, *Mimetus* Zone, Late Anisian.

13a–b: ***Tropigymnites* sp. indet.** PIMUZ 25456.
Loc. HB 592, Oliver Gulch (Augusta Mountains). *Cordeyi* Subzone, *Weitschati* Zone, Late Anisian.

14a–c: ***Tropigymnites* sp. indet.** PIMUZ 25457.
Loc. HB 713, Oliver Gulch (Augusta Mountains). *Cordeyi* Subzone, *Weitschati* Zone, Late Anisian.

PLATE 17

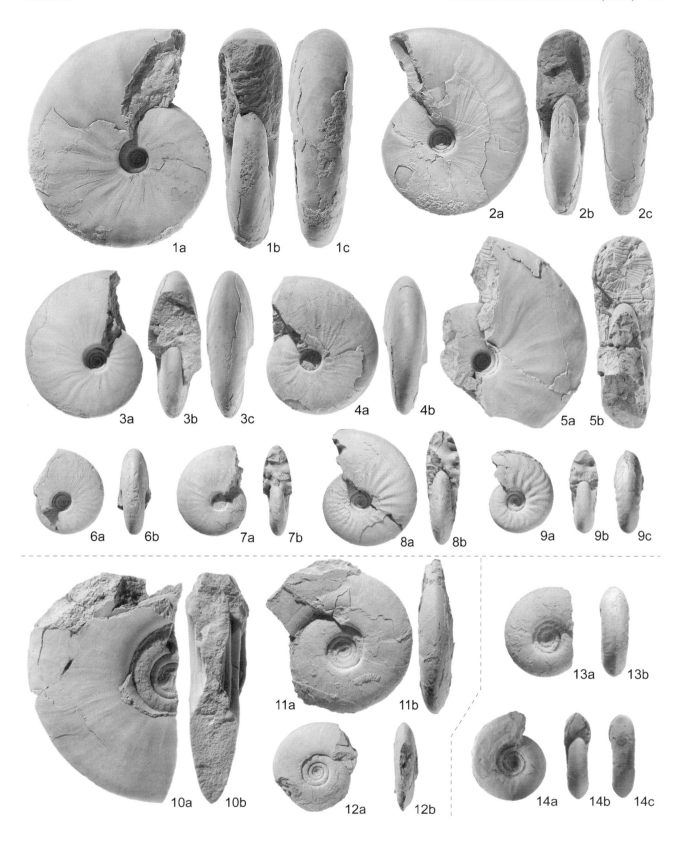

Plate 18 (all figures natural size)

1a–c: ***Dixieceras lawsoni* (Smith, 1914).** PIMUZ 25252.
Loc. FHB 8, Fossil Hill (Humboldt Range). *Lawsoni* Subzone, *Mimetus* Zone, Late Anisian. Injured specimen developing a ventral ribbing.

2a–b: ***Dixieceras lawsoni* (Smith, 1914).** PIMUZ 25253.
Loc. FHB 9, Fossil Hill (Humboldt Range). *Lawsoni* Subzone, *Mimetus* Zone, Late Anisian.

3a–c: ***Sageceras walteri* Mojsisovics, 1882.** PIMUZ 25424.
Loc. HB 2007, Rieber Gulch (Augusta Mountains). *Cordeyi* Subzone, *Weitschati* Zone, Late Anisian.

4a–c: ***Sageceras walteri* Mojsisovics, 1882.** PIMUZ 25425.
Loc. HB 2041, Oliver Gulch (Augusta Mountains). *Vogdesi* Subzone, *Rotelliformis* Zone, Late Anisian.

5a–b: ***Sageceras walteri* Mojsisovics, 1882.** PIMUZ 25426.
Loc. FHB 8, Fossil Hill (Humboldt Range). *Lawsoni* Subzone, *Mimetus* Zone, Late Anisian.

6a–c: ***Gymnotoceras blakei* (Gabb, 1864).** PIMUZ 25265.
Loc. FHB 30, Fossil Hill (Humboldt Range). *Balkei* Subzone, *Rotelliformis* Zone, Late Anisian.

7a–b: ***Gymnotoceras blakei* (Gabb, 1864).** PIMUZ 25266.
Loc. FHB 34, Fossil Hill (Humboldt Range). *Blakei* Subzone, *Rotelliformis* Zone, Late Anisian.

PLATE 18

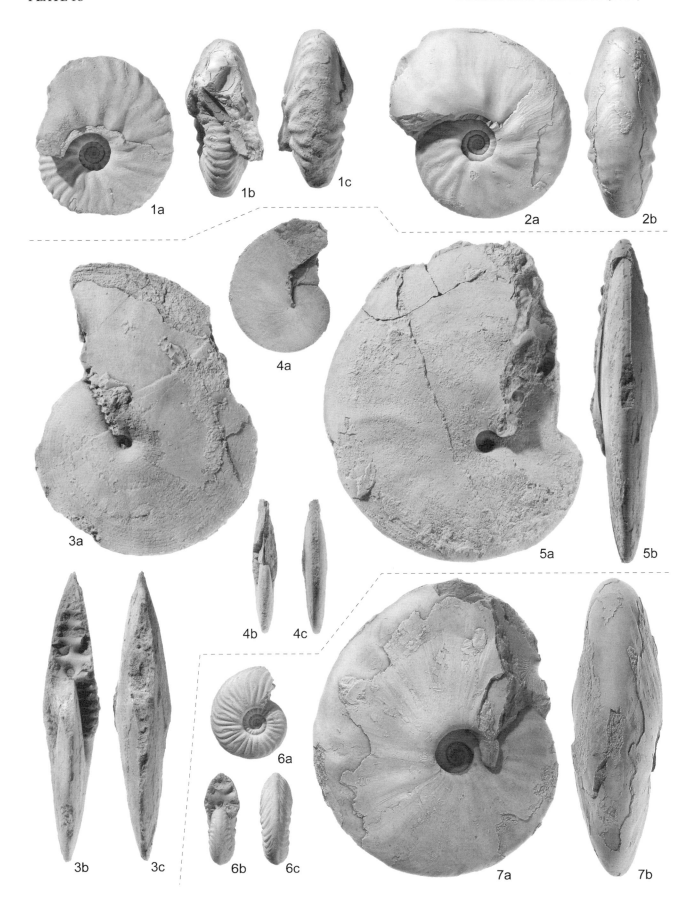

Plate 19 (all figures natural size)

1a–c: *Jenksites flexicostatus* **n. gen. n. sp.** PIMUZ 25321, holotype. Loc. HB 598, Oliver Gulch (Augusta Mountains). *Cordeyi* Subzone, *Weitschati* Zone, Late Anisian.

2a–b: *Jenksites flexicostatus* **n. gen. n. sp.** PIMUZ 25323. Loc. HB 713, Oliver Gulch (Augusta Mountains). *Cordeyi* Subzone, *Weitschati* Zone, Late Anisian.

3a–c: *Jenksites flexicostatus* **n. gen. n. sp.** PIMUZ 25322, paratype. Loc. HB 598, Oliver Gulch (Augusta Mountains). *Cordeyi* Subzone, *Weitschati* Zone, Late Anisian.

4a–c: *Jenksites flexicostatus* **n. gen. n. sp.** PIMUZ 25324. Loc. HB 713, Oliver Gulch (Augusta Mountains). *Cordeyi* Subzone, *Weitschati* Zone, Late Anisian.

5a–b: *Jenksites flexicostatus* **n. gen. n. sp.** PIMUZ 25327. Loc. HB 2007, Rieber Gulch (Augusta Mountains). *Cordeyi* Subzone, *Weitschati* Zone, Late Anisian.

6a–b: *Jenksites flexicostatus* **n. gen. n. sp.** PIMUZ 25325. Loc. HB 713, Oliver Gulch (Augusta Mountains). *Cordeyi* Subzone, *Weitschati* Zone, Late Anisian.

7a–c: *Jenksites flexicostatus* **n. gen. n. sp.** PIMUZ 25326. Loc. HB 713, Oliver Gulch (Augusta Mountains). *Cordeyi* Subzone, *Weitschati* Zone, Late Anisian.

8a–b: *Jenksites flexicostatus* **n. gen. n. sp.** PIMUZ 25328. Loc. HB 2007, Rieber Gulch (Augusta Mountains). *Cordeyi* Subzone, *Weitschati* Zone, Late Anisian.

9a–c: *Jenksites flexicostatus* **n. gen. n. sp.** PIMUZ 25329. Loc. HB 2007, Rieber Gulch (Augusta Mountains). *Cordeyi* Subzone, *Weitschati* Zone, Late Anisian.

10a–b: *Jenksites flexicostatus* **n. gen. n. sp.** PIMUZ 25330. Loc. HB 2007, Rieber Gulch (Augusta Mountains). *Cordeyi* Subzone, *Weitschati* Zone, Late Anisian.

11a–b: *Jenksites flexicostatus* **n. gen. n. sp.** PIMUZ 25331. Loc. HB 2007, Rieber Gulch (Augusta Mountains). *Cordeyi* Subzone, *Weitschati* Zone, Late Anisian.

12a–b: *Jenksites flexicostatus* **n. gen. n. sp.** PIMUZ 25332. Loc. HB 2007, Rieber Gulch (Augusta Mountains). *Cordeyi* Subzone, *Weitschati* Zone, Late Anisian.

13a–b: *Jenksites flexicostatus* **n. gen. n. sp.** PIMUZ 25333. Loc. HB 2007, Rieber Gulch (Augusta Mountains). *Cordeyi* Subzone, *Weitschati* Zone, Late Anisian.

14a–b: *Jenksites flexicostatus* **n. gen. n. sp.** PIMUZ 25334. Loc. HB 2007, Rieber Gulch (Augusta Mountains). *Cordeyi* Subzone, *Weitschati* Zone, Late Anisian.

15a–c: *Jenksites flexicostatus* **n. gen. n. sp.** PIMUZ 25335. Loc. HB 2007, Rieber Gulch (Augusta Mountains). *Cordeyi* Subzone, *Weitschati* Zone, Late Anisian.

PLATE 19

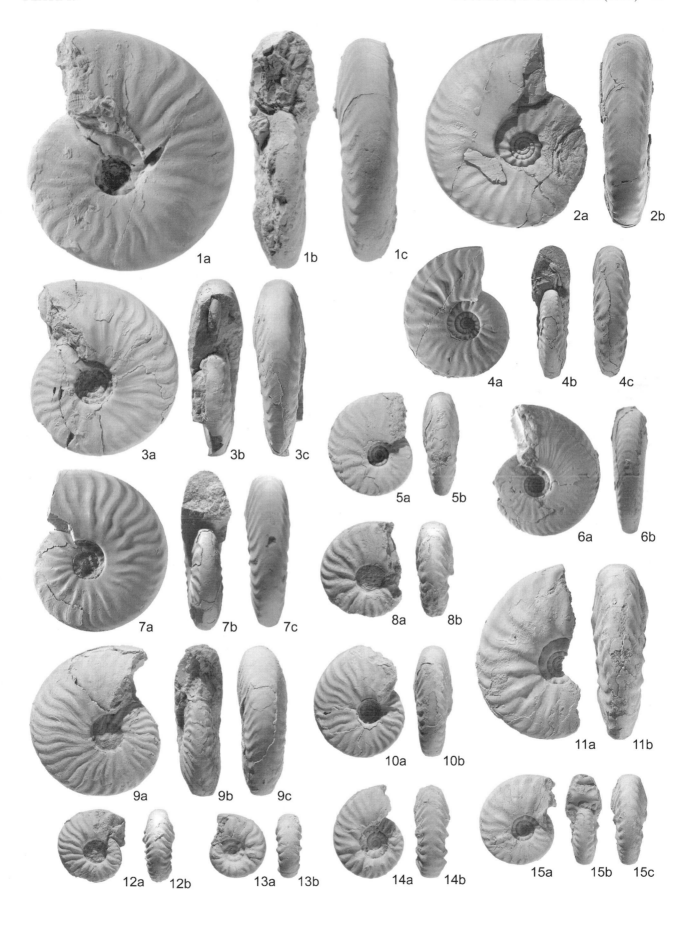

Plate 20 (all figures natural size)

1a–c: ***Rieppelites shevyrevi* n. gen. n. sp.** PIMUZ 25392.
Loc. HB 713, Oliver Gulch (Augusta Mountains). *Cordeyi*
Subzone, *Weitschati* Zone, Late Anisian.

2a–c: ***Rieppelites shevyrevi* n. gen. n. sp.** PIMUZ 25393.
Loc. HB 713, Oliver Gulch (Augusta Mountains). *Cordeyi*
Subzone, *Weitschati* Zone, Late Anisian.

3a–b: ***Rieppelites shevyrevi* n. gen. n. sp.** PIMUZ 25394.
Loc. HB 598, Oliver Gulch (Augusta Mountains). *Cordeyi*
Subzone, *Weitschati* Zone, Late Anisian.

4a–b: ***Rieppelites shevyrevi* n. gen. n. sp.** PIMUZ 25395.
Loc. HB 734, Muller Canyon (Augusta Mountains). *Cordeyi*
Subzone, *Weitschati* Zone, Late Anisian.

5a–b: ***Rieppelites shevyrevi* n. gen. n. sp.** PIMUZ 25396.
Loc. HB 734, Muller Canyon (Augusta Mountains). *Cordeyi*
Subzone, *Weitschati* Zone, Late Anisian.

6a–b: ***Rieppelites shevyrevi* n. gen. n. sp.** PIMUZ 25403.
Loc. JJ 1997/2, Favret Canyon (Augusta Mountains). *Cordeyi*
Subzone, *Weitschati* Zone, Late Anisian.

7a–b: ***Rieppelites shevyrevi* n. gen. n. sp.** PIMUZ 25397.
Loc. HB 592, Oliver Gulch (Augusta Mountains). *Cordeyi*
Subzone, *Weitschati* Zone, Late Anisian.

8a–b: ***Rieppelites shevyrevi* n. gen. n. sp.** PIMUZ 25398.
Loc. HB 593, Rieber Gulch (Augusta Mountains). *Cordeyi*
Subzone, *Weitschati* Zone, Late Anisian.

9a–b: ***Rieppelites shevyrevi* n. gen. n. sp.** PIMUZ 25399.
Loc. HB 593, Oliver Gulch (Augusta Mountains). *Cordeyi*
Subzone, *Weitschati* Zone, Late Anisian.

10a–b: ***Rieppelites shevyrevi* n. gen. n. sp.** PIMUZ 25400.
Loc. HB 731, Muller Canyon (Augusta Mountains). *Cordeyi*
Subzone, *Weitschati* Zone, Late Anisian.

11a–c: ***Rieppelites shevyrevi* n. gen. n. sp.** PIMUZ 25401.
Loc. JJ 1997/2, Favret Canyon (Augusta Mountains). *Cordeyi*
Subzone, *Weitschati* Zone, Late Anisian.

12a–c: ***Rieppelites shevyrevi* n. gen. n. sp.** PIMUZ 25402.
Loc. JJ 1997/2, Favret Canyon (Augusta Mountains). *Cordeyi*
Subzone, *Weitschati* Zone, Late Anisian.

PLATE 20

FOSSILS AND STRATA 52 (2005)   99

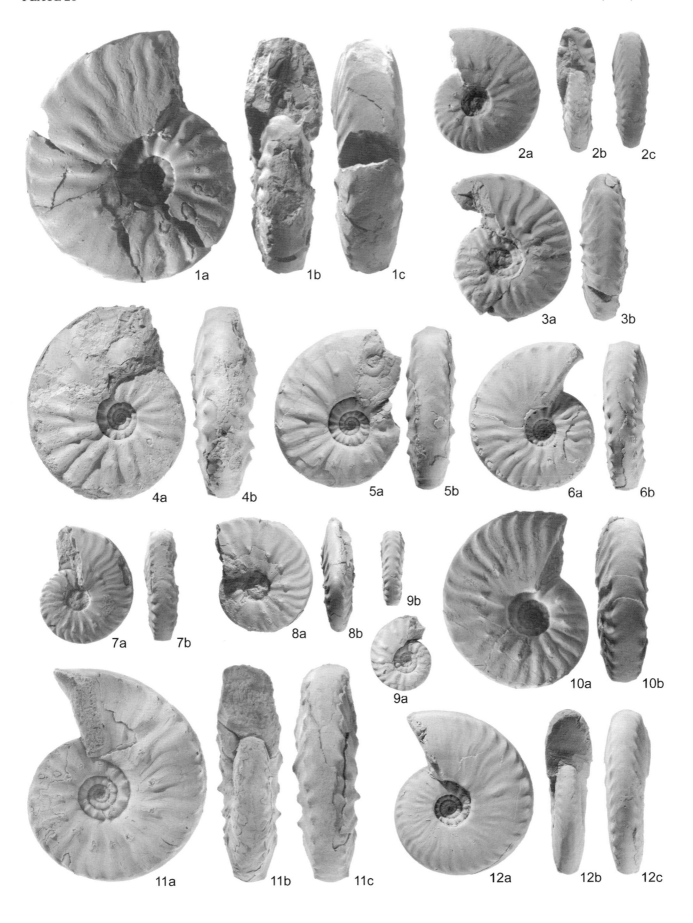

Plate 21 (all figures natural size)

1a–b: ***Rieppelites shevyrevi* n. gen. n. sp.** PIMUZ 25404, paratype. Loc. HB 2030, Ferguson West (Augusta Mountains). *Cordeyi* Subzone, *Weitschati* Zone, Late Anisian.

2a–b: ***Rieppelites shevyrevi* n. gen. n. sp.** PIMUZ 25405, holotype. Loc. HB 2030, Ferguson West (Augusta Mountains). *Cordeyi* Subzone, *Weitschati* Zone, Late Anisian.

3a–b: ***Rieppelites shevyrevi* n. gen. n. sp.** PIMUZ 25406, paratype. Loc. HB 2030, Ferguson West (Augusta Mountains). *Cordeyi* Subzone, *Weitschati* Zone, Late Anisian.

4a–b: ***Rieppelites shevyrevi* n. gen. n. sp.** PIMUZ 25407, paratype. Loc. HB 2030, Ferguson West (Augusta Mountains). *Cordeyi* Subzone, *Weitschati* Zone, Late Anisian.

5a–c: ***Rieppelites shevyrevi* n. gen. n. sp.** PIMUZ 25408, paratype. Loc. HB 2030, Ferguson West (Augusta Mountains). *Cordeyi* Subzone, *Weitschati* Zone, Late Anisian.

6a–c: ***Rieppelites shevyrevi* n. gen. n. sp.** PIMUZ 25409, paratype. Loc. HB 2030, Ferguson West (Augusta Mountains). *Cordeyi* Subzone, *Weitschati* Zone, Late Anisian.

7a–b: ***Rieppelites shevyrevi* n. gen. n. sp.** PIMUZ 25410, paratype. Loc. HB 2030, Ferguson West (Augusta Mountains). *Cordeyi* Subzone, *Weitschati* Zone, Late Anisian.

8a–b: ***Rieppelites shevyrevi* n. gen. n. sp.** PIMUZ 25411, paratype. Loc. HB 2030, Ferguson West (Augusta Mountains). *Cordeyi* Subzone, *Weitschati* Zone, Late Anisian.

9a–c: ***Rieppelites shevyrevi* n. gen. n. sp.** PIMUZ 25412, paratype. Loc. HB 2030, Ferguson West (Augusta Mountains). *Cordeyi* Subzone, *Weitschati* Zone, Late Anisian.

10a–c: ***Rieppelites shevyrevi* n. gen. n. sp.** PIMUZ 25413, paratype. Loc. HB 2030, Ferguson West (Augusta Mountains). *Cordeyi* Subzone, *Weitschati* Zone, Late Anisian. Injured specimen showing lateral compensation of ornamentation.

11a–b: ***Rieppelites shevyrevi* n. gen. n. sp.** PIMUZ 25414, paratype. Loc. HB 2030, Ferguson West (Augusta Mountains). *Cordeyi* Subzone, *Weitschati* Zone, Late Anisian. Specimen showing teratological ventral ribs.

12a–b: ***Rieppelites shevyrevi* n. gen. n. sp.** PIMUZ 25415. Loc. HB 2062, McCoy Mine (New Pass Range). *Cordeyi* Subzone, *Weitschati* Zone, Late Anisian.

13a–c: ***Rieppelites shevyrevi* n. gen. n. sp.** PIMUZ 25419. Loc. HB 2010, Rieber Gulch (Augusta Mountains). *Cordeyi* Subzone, *Weitschati* Zone, Late Anisian.

14a–c: ***Rieppelites shevyrevi* n. gen. n. sp.** PIMUZ 25416. Loc. HB 2062, McCoy Mine (New Pass Range). *Cordeyi* Subzone, *Weitschati* Zone, Late Anisian.

15a–c: ***Rieppelites shevyrevi* n. gen. n. sp.** PIMUZ 25417. Loc. HB 2062, McCoy Mine (New Pass Range). *Cordeyi* Subzone, *Weitschati* Zone, Late Anisian.

16a–c: ***Rieppelites shevyrevi* n. gen. n. sp.** PIMUZ 25420. Loc. HB 2010, Rieber Gulch (Augusta Mountains). *Cordeyi* Subzone, *Weitschati* Zone, Late Anisian.

17a–b: ***Rieppelites shevyrevi* n. gen. n. sp.** PIMUZ 25421. Loc. HB 2010, Rieber Gulch (Augusta Mountains). *Cordeyi* Subzone, *Weitschati* Zone, Late Anisian.

18a–c: ***Rieppelites shevyrevi* n. gen. n. sp.** PIMUZ 25418. Loc. HB 2062, McCoy Mine (New Pass Range). *Cordeyi* Subzone, *Weitschati* Zone, Late Anisian.

PLATE 21

FOSSILS AND STRATA 52 (2005)   101

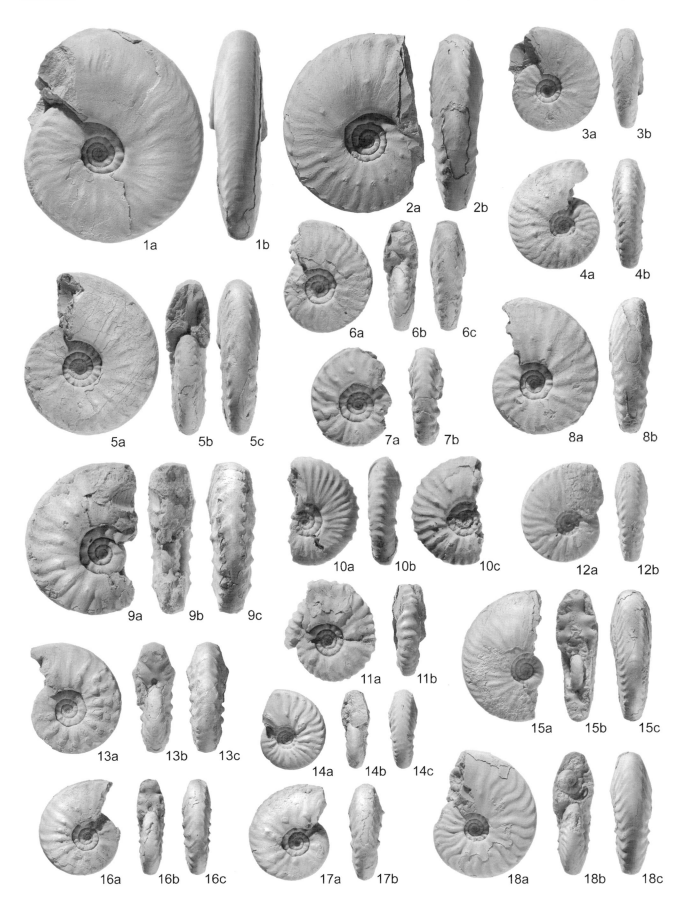

1a   1b

2a   2b

3a   3b

4a   4b

5a   5b   5c

6a   6b   6c

7a   7b

8a   8b

9a   9b   9c

10a   10b   10c

12a   12b

11a   11b

13a   13b   13c

14a   14b   14c

15a   15b   15c

16a   16b   16c

17a   17b

18a   18b   18c

Plate 22 (all figures natural size)

1a–b: ***Rieppelites boletzkyi* n. gen. n. sp.** PIMUZ 25372, holotype. Loc. HB 2006, Rieber Gulch (Augusta Mountains). *Cordeyi* Subzone, *Weitschati* Zone, Late Anisian.

2a–b: ***Rieppelites boletzkyi* n. gen. n. sp.** PIMUZ 25373, paratype. Loc. HB 2006, Rieber Gulch (Augusta Mountains). *Cordeyi* Subzone, *Weitschati* Zone, Late Anisian.

3a–b: ***Rieppelites boletzkyi* n. gen. n. sp.** PIMUZ 25374, paratype. Loc. HB 2006, Rieber Gulch (Augusta Mountains). *Cordeyi* Subzone, *Weitschati* Zone, Late Anisian.

4a–b: ***Rieppelites boletzkyi* n. gen. n. sp.** PIMUZ 25375, paratype. Loc. HB 2006, Rieber Gulch (Augusta Mountains). *Cordeyi* Subzone, *Weitschati* Zone, Late Anisian.

5a–b: ***Rieppelites boletzkyi* n. gen. n. sp.** PIMUZ 25376, paratype. Loc. HB 2006, Rieber Gulch (Augusta Mountains). *Cordeyi* Subzone, *Weitschati* Zone, Late Anisian.

6a–c: ***Rieppelites boletzkyi* n. gen. n. sp.** PIMUZ 25377, paratype. Loc. HB 2006, Rieber Gulch (Augusta Mountains). *Cordeyi* Subzone, *Weitschati* Zone, Late Anisian.

7a–c: ***Rieppelites boletzkyi* n. gen. n. sp.** PIMUZ 25378, paratype. Loc. HB 2006, Rieber Gulch (Augusta Mountains). *Cordeyi* Subzone, *Weitschati* Zone, Late Anisian.

8a–c: ***Rieppelites boletzkyi* n. gen. n. sp.** PIMUZ 25379, paratype. Loc. HB 2006, Rieber Gulch (Augusta Mountains). *Cordeyi* Subzone, *Weitschati* Zone, Late Anisian.

9a–b: ***Rieppelites boletzkyi* n. gen. n. sp.** PIMUZ 25382. Loc. HB 2009, Rieber Gulch (Augusta Mountains). *Cordeyi* Subzone, *Weitschati* Zone, Late Anisian.

10a–b: ***Rieppelites boletzkyi* n. gen. n. sp.** PIMUZ 25380, paratype. Loc. HB 2006, Rieber Gulch (Augusta Mountains). *Cordeyi* Subzone, *Weitschati* Zone, Late Anisian.

11a–b: ***Rieppelites boletzkyi* n. gen. n. sp.** PIMUZ 25381, paratype. Loc. HB 2006, Rieber Gulch (Augusta Mountains). *Cordeyi* Subzone, *Weitschati* Zone, Late Anisian.

PLATE 22

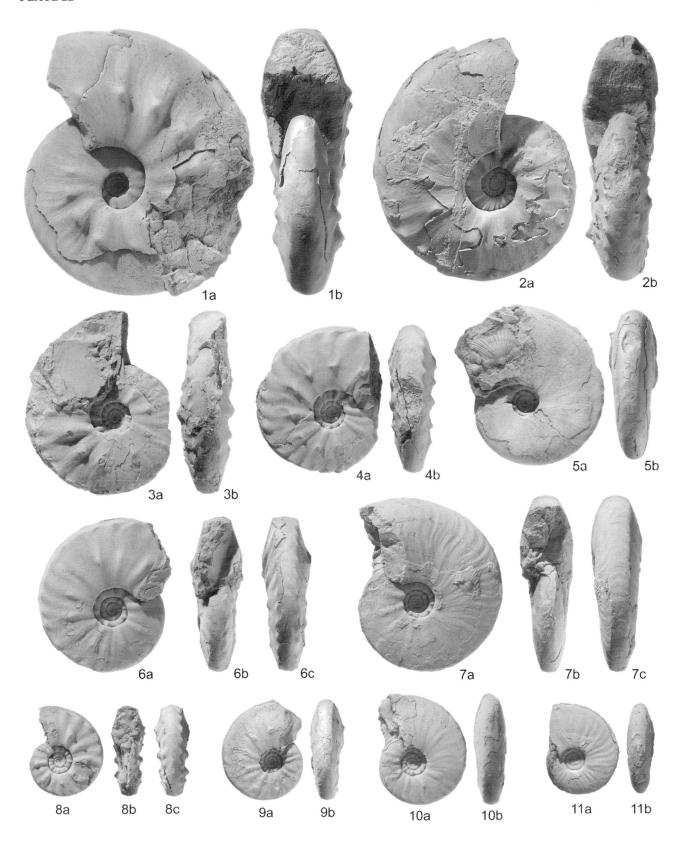

Plate 23 (all figures natural size)

1a–b: *Rieppelites boletzkyi* **n. gen. n. sp.** PIMUZ 25383, paratype.
Loc. HB 2006, Rieber Gulch (Augusta Mountains). *Cordeyi*
Subzone, *Weitschati* Zone, Late Anisian.

2a–c: *Rieppelites boletzkyi* **n. gen. n. sp.** PIMUZ 25384, paratype.
Loc. HB 2006, Rieber Gulch (Augusta Mountains). *Cordeyi*
Subzone, *Weitschati* Zone, Late Anisian.

3a–b: *Rieppelites boletzkyi* **n. gen. n. sp.** PIMUZ 25390.
Loc. HB 713, Oliver Gulch (Augusta Mountains). *Cordeyi*
Subzone, *Weitschati* Zone, Late Anisian.

4a–c: *Rieppelites boletzkyi* **n. gen. n. sp.** PIMUZ 25385, paratype.
Loc. HB 2006, Rieber Gulch (Augusta Mountains). *Cordeyi*
Subzone, *Weitschati* Zone, Late Anisian.

5a–b: *Rieppelites boletzkyi* **n. gen. n. sp.** PIMUZ 25386, paratype.
Loc. HB 2006, Rieber Gulch (Augusta Mountains). *Cordeyi*
Subzone, *Weitschati* Zone, Late Anisian.

6a–c: *Rieppelites boletzkyi* **n. gen. n. sp.** PIMUZ 25387, paratype.
Loc. HB 2006, Rieber Gulch (Augusta Mountains). *Cordeyi*
Subzone, *Weitschati* Zone, Late Anisian.

7a–b: *Rieppelites boletzkyi* **n. gen. n. sp.** PIMUZ 25388, paratype.
Loc. HB 2006, Rieber Gulch (Augusta Mountains). *Cordeyi*
Subzone, *Weitschati* Zone, Late Anisian.

8a–b: *Rieppelites boletzkyi* **n. gen. n. sp.** PIMUZ 25389, paratype.
Loc. HB 2006, Rieber Gulch (Augusta Mountains). *Cordeyi*
Subzone, *Weitschati* Zone, Late Anisian.

9a–c: *Ptychites* **sp. indet.** PIMUZ 25361.
Loc. HB 2001, McCoy Mine (New Pass Range). *Cordeyi*
Subzone, *Weitschati* Zone, Late Anisian. Arrow indicates the
location of a varice.

10a–b: *Discoptychites* **cf.** *D. megalodiscus* **(Beyrich, 1867).** PIMUZ
25218.
Loc. HB 596, Oliver Gulch (Augusta Mountains). *Transiformis*
Subzone, *Weitschati* Zone, Late Anisian.

11a–c: *Ptychites* **sp. indet.** PIMUZ 25360.
Loc. HB 2001, McCoy Mine (New Pass Range). *Cordeyi*
Subzone, *Weitschati* Zone, Late Anisian.

12a–c: *Ptychites* **sp. indet.** PIMUZ 25362.
Loc. HB 2007, Rieber Gulch (Augusta Mountains). *Cordeyi*
Subzone, *Weitschati* Zone, Late Anisian.

PLATE 23

FOSSILS AND STRATA 52 (2005)   105

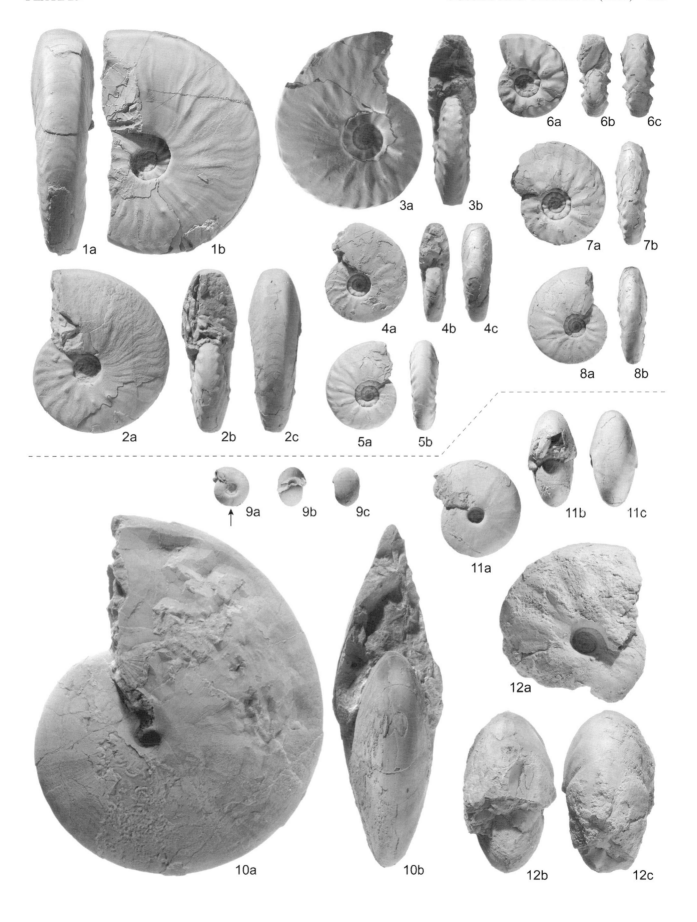

1a   1b

2a   2b   2c

3a   3b

4a   4b   4c

5a   5b

6a   6b   6c

7a   7b

8a   8b

9a   9b   9c

10a   10b

11a   11b   11c

12a   12b   12c

Plate 24 (all figures natural size)

1a–b: ***Rieberites transiformis* n. gen. n. sp.** PIMUZ 25363, paratype. Loc. HB 596, Oliver Gulch (Augusta Mountains). *Transiformis* Subzone, *Weitschati* Zone, Late Anisian.

2a–c: ***Rieberites transiformis* n. gen. n. sp.** PIMUZ 25364, paratype. Loc. HB 596, Oliver Gulch (Augusta Mountains). *Transiformis* Subzone, *Weitschati* Zone, Late Anisian.

3a–b: ***Rieberites transiformis* n. gen. n. sp.** PIMUZ 25365, paratype. Loc. HB 596, Oliver Gulch (Augusta Mountains). *Transiformis* Subzone, *Weitschati* Zone, Late Anisian.

4a–b: ***Rieberites transiformis* n. gen. n. sp.** PIMUZ 25366, holotype. Loc. HB 596, Oliver Gulch (Augusta Mountains). *Transiformis* Subzone, *Weitschati* Zone, Late Anisian.

5a–c: ***Rieberites transiformis* n. gen. n. sp.** PIMUZ 25367, paratype. Loc. HB 596, Oliver Gulch (Augusta Mountains). *Transiformis* Subzone, *Weitschati* Zone, Late Anisian.

6a–c: ***Rieberites transiformis* n. gen. n. sp.** PIMUZ 25368, paratype. Loc. HB 596, Oliver Gulch (Augusta Mountains). *Transiformis* Subzone, *Weitschati* Zone, Late Anisian.

7a–c: ***Rieberites transiformis* n. gen. n. sp.** PIMUZ 25369, paratype. Loc. HB 596, Oliver Gulch (Augusta Mountains). *Transiformis* Subzone, *Weitschati* Zone, Late Anisian.

PLATE 24

Plate 25 (all figures natural size)

1a–c: ***Rieberites transiformis* n. gen. n. sp.** PIMUZ 25370, paratype.
Loc. HB 596, Oliver Gulch (Augusta Mountains). *Transiformis* Subzone, *Weitschati* Zone, Late Anisian.

2a–b: ***Rieberites transiformis* n. gen. n. sp.** PIMUZ 25371, paratype.
Loc. HB 596, Oliver Gulch (Augusta Mountains). *Transiformis* Subzone, *Weitschati* Zone, Late Anisian.

3a–b: ***Silberlingia praecursor* n. gen. n. sp.** PIMUZ 25449, holotype.
Loc. HB 596, Oliver Gulch (Augusta Mountains). *Transiformis* Subzone, *Weitschati* Zone, Late Anisian.

4a–b: ***Silberlingia praecursor* n. gen. n. sp.** PIMUZ 25450, paratype.
Loc. HB 596, Oliver Gulch (Augusta Mountains). *Transiformis* Subzone, *Weitschati* Zone, Late Anisian. Teratological ornamentation resulting from shell breakage (arrow).

5a–c: ***Silberlingia cricki* (Smith, 1914).** PIMUZ 25428.
Loc. FHB 22, Fossil Hill (Humboldt Range). *Vogdesi* Subzone, *Rotelliformis* Zone, Late Anisian. Teratological specimen.

PLATE 25

FOSSILS AND STRATA 52 (2005)   109

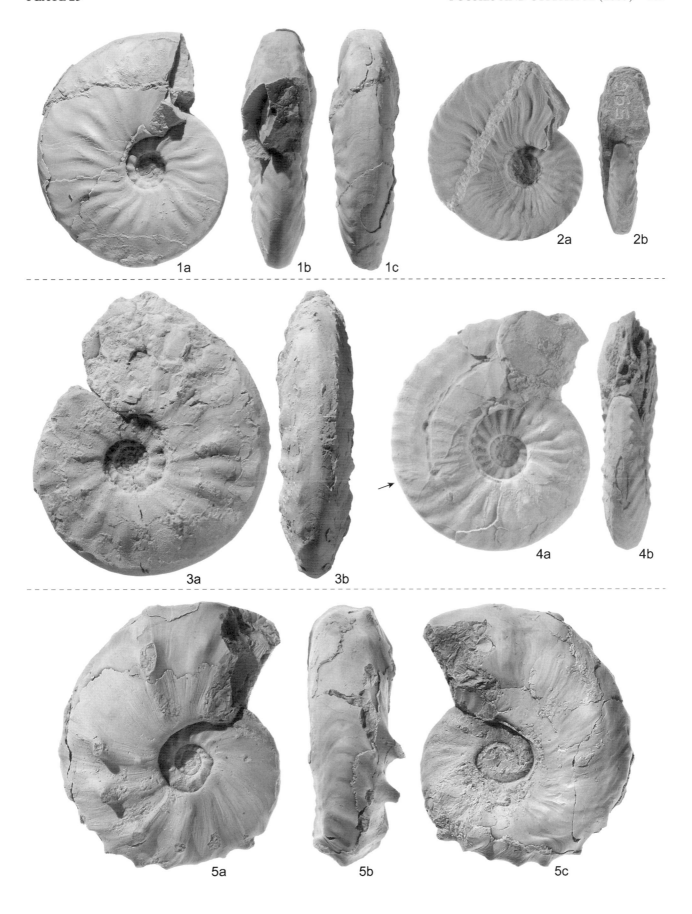

1a   1b   1c   2a   2b

3a   3b   4a   4b

5a   5b   5c

Plate 26 (all figures natural size)

1a–c: ***Silberlingia cricki* (Smith, 1914).** PIMUZ 25429.
Loc. FHB 15a, Fossil Hill (Humboldt Range). *Vogdesi* Subzone, *Rotelliformis* Zone, Late Anisian.

2a–b: ***Silberlingia cricki* (Smith, 1914).** PIMUZ 25436.
Loc. FHB 22, Fossil Hill (Humboldt Range). *Vogdesi* Subzone, *Rotelliformis* Zone, Late Anisian.

3a–b: ***Silberlingia cricki* (Smith, 1914).** PIMUZ 25430.
Loc. FHB 15a, Fossil Hill (Humboldt Range). *Vogdesi* Subzone, *Rotelliformis* Zone, Late Anisian.

4a–b: ***Silberlingia cricki* (Smith, 1914).** PIMUZ 25431.
Loc. FHB 15a, Fossil Hill (Humboldt Range). *Vogdesi* Subzone, *Rotelliformis* Zone, Late Anisian.

5a–c: ***Silberlingia cricki* (Smith, 1914).** PIMUZ 25432.
Loc. FHB 15a, Fossil Hill (Humboldt Range). *Vogdesi* Subzone, *Rotelliformis* Zone, Late Anisian.

6a–b: ***Silberlingia cricki* (Smith, 1914).** PIMUZ 25433.
Loc. FHB 15a, Fossil Hill (Humboldt Range). *Vogdesi* Subzone, *Rotelliformis* Zone, Late Anisian.

7a–b: ***Silberlingia cricki* (Smith, 1914).** PIMUZ 25434.
Loc. FHB 15a, Fossil Hill (Humboldt Range). *Vogdesi* Subzone, *Rotelliformis* Zone, Late Anisian.

8a–b: ***Silberlingia cricki* (Smith, 1914).** PIMUZ 25435.
Loc. FHB 22, Fossil Hill (Humboldt Range). *Vogdesi* Subzone, *Rotelliformis* Zone, Late Anisian.

PLATE 26

FOSSILS AND STRATA 52 (2005) 111

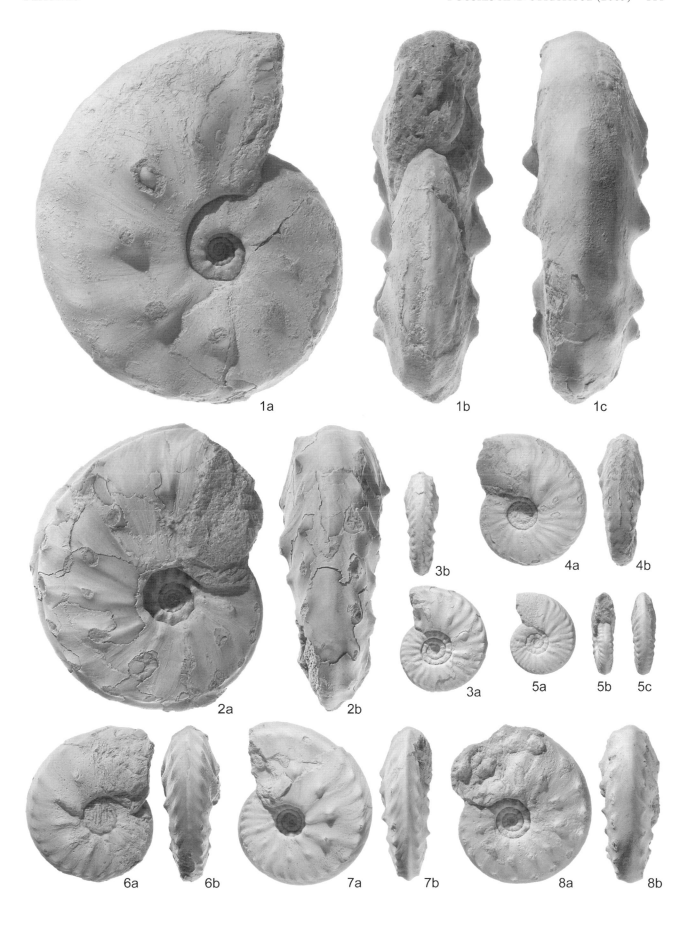

1a

1b

1c

2a

2b

3b

3a

4a

4b

5a

5b

5c

6a

6b

7a

7b

8a

8b

Plate 27 (all figures natural size)

1a–c: *Silberlingia cricki* **(Smith, 1914).** PIMUZ 25437.
Loc. FHB 15a, Fossil Hill (Humboldt Range). *Vogdesi* Subzone, *Rotelliformis* Zone, Late Anisian.

2a–b: *Silberlingia cricki* **(Smith, 1914).** PIMUZ 25438.
Loc. FHB 15a, Fossil Hill (Humboldt Range). *Vogdesi* Subzone, *Rotelliformis* Zone, Late Anisian.

3a–b: *Silberlingia cricki* **(Smith, 1914).** PIMUZ 25439.
Loc. FHB 15a, Fossil Hill (Humboldt Range). *Vogdesi* Subzone, *Rotelliformis* Zone, Late Anisian.

4a–b: *Silberlingia cricki* **(Smith, 1914).** PIMUZ 25440.
Loc. FHB 15a, Fossil Hill (Humboldt Range). *Vogdesi* Subzone, *Rotelliformis* Zone, Late Anisian.

5a–b: *Silberlingia cricki* **(Smith, 1914).** PIMUZ 25445.
Loc. IIB 702, Oliver Gulch (Augusta Mountains). *Rotelliformis* Zone, Late Anisian.

6a–c: *Silberlingia cricki* **(Smith, 1914).** PIMUZ 25444.
Loc. HB 702, Oliver Gulch (Augusta Mountains). *Rotelliformis* Zone, Late Anisian.

7a–b: *Silberlingia cricki* **(Smith, 1914).** PIMUZ 25441.
Loc. FHB 13, Fossil Hill (Humboldt Range). *Vogdesi* Subzone, *Rotelliformis* Zone, Late Anisian.

8a–c: *Silberlingia cricki* **(Smith, 1914).** PIMUZ 25443.
Loc. HB 747, Oliver Gulch (Augusta Mountains). *Vogdesi* Subzone, *Rotelliformis* Zone, Late Anisian.

9a–c: *Silberlingia cricki* **(Smith, 1914).** PIMUZ 25442.
Loc. FHB 13, Fossil Hill (Humboldt Range). *Vogdesi* Subzone, *Rotelliformis* Zone, Late Anisian.

10a–b: *Silberlingia cricki* **(Smith, 1914).** PIMUZ 25447.
Loc. HB 702, Oliver Gulch (Augusta Mountains). *Rotelliformis* Zone, Late Anisian.

11a–c: *Silberlingia cricki* **(Smith, 1914).** PIMUZ 25446.
Loc. HB 702, Oliver Gulch (Augusta Mountains). *Rotelliformis* Zone, Late Anisian.

12a–c: *Silberlingia cricki* **(Smith, 1914).** PIMUZ 25448.
Loc. HB 702, Oliver Gulch (Augusta Mountains). *Rotelliformis* Zone, Late Anisian.

PLATE 27

FOSSILS AND STRATA 52 (2005)   113

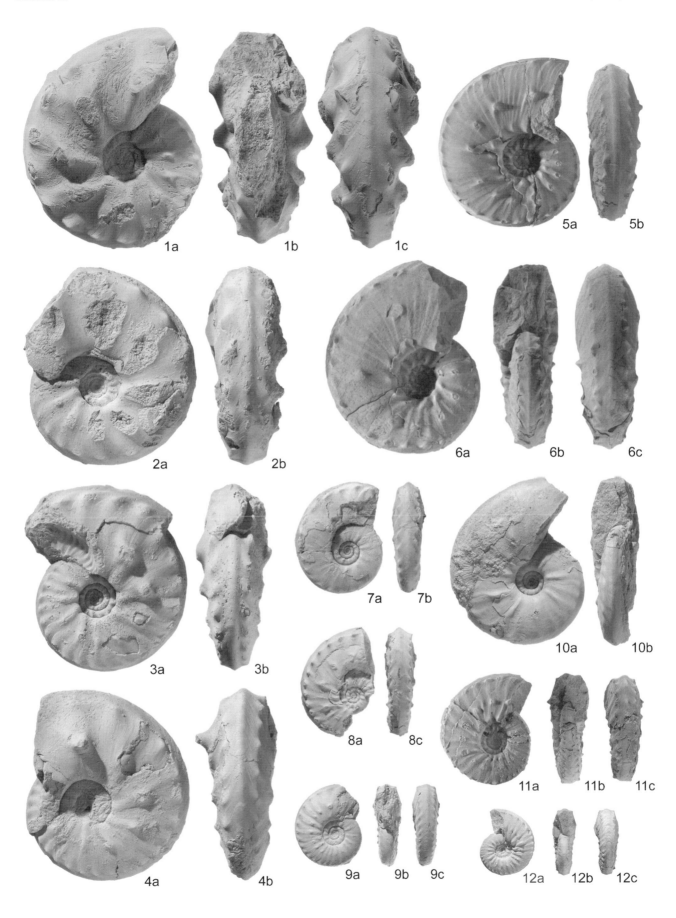

Plate 28 (all figures natural size)

1a–c: ***Brackites vogdesi* (Smith, 1904).** PIMUZ 25184.
Loc. JJ 2002/1, Favret Canyon (Augusta Mountains). *Vogdesi* Subzone, *Rotelliformis* Zone, Late Anisian.

2a–b: ***Brackites vogdesi* (Smith, 1904).** PIMUZ 25185.
Loc. HB 2047, Favret Canyon (Augusta Mountains). *Vogdesi* Subzone, *Rotelliformis* Zone, Late Anisian.

3a–b: ***Brackites vogdesi* (Smith, 1904).** PIMUZ 25186.
Loc. HB 745, Oliver Gulch (Augusta Mountains). *Vogdesi* Subzone, *Rotelliformis* Zone, Late Anisian.

4a–b: ***Brackites vogdesi* (Smith, 1904).** PIMUZ 25187.
Loc. HB 745, Oliver Gulch (Augusta Mountains). *Vogdesi* Subzone, *Rotelliformis* Zone, Late Anisian.

5a–c: ***Brackites vogdesi* (Smith, 1904).** PIMUZ 25188.
Loc. HB 745, Oliver Gulch (Augusta Mountains). *Vogdesi* Subzone, *Rotelliformis* Zone, Late Anisian. Specimen showing teratological ventral ribs.

6a–c: ***Brackites vogdesi* (Smith, 1904).** PIMUZ 25189.
Loc. HB 742, Oliver Gulch (Augusta Mountains). *Vogdesi* Subzone, *Rotelliformis* Zone, Late Anisian.

7a–b: ***Brackites vogdesi* (Smith, 1904).** PIMUZ 25190.
Loc. HB 742, Oliver Gulch (Augusta Mountains). *Vogdesi* Subzone, *Rotelliformis* Zone, Late Anisian.

8a–b: ***Brackites vogdesi* (Smith, 1904).** PIMUZ 25191.
Loc. HB 2063, McCoy Mine (New Pass Range). *Vogdesi* Subzone, *Rotelliformis* Zone, Late Anisian.

9a–b: ***Brackites vogdesi* (Smith, 1904).** PIMUZ 25192.
Loc. HB 2047, Favret Canyon (Augusta Mountains). *Vogdesi* Subzone, *Rotelliformis* Zone, Late Anisian.

PLATE 28

FOSSILS AND STRATA 52 (2005)   115

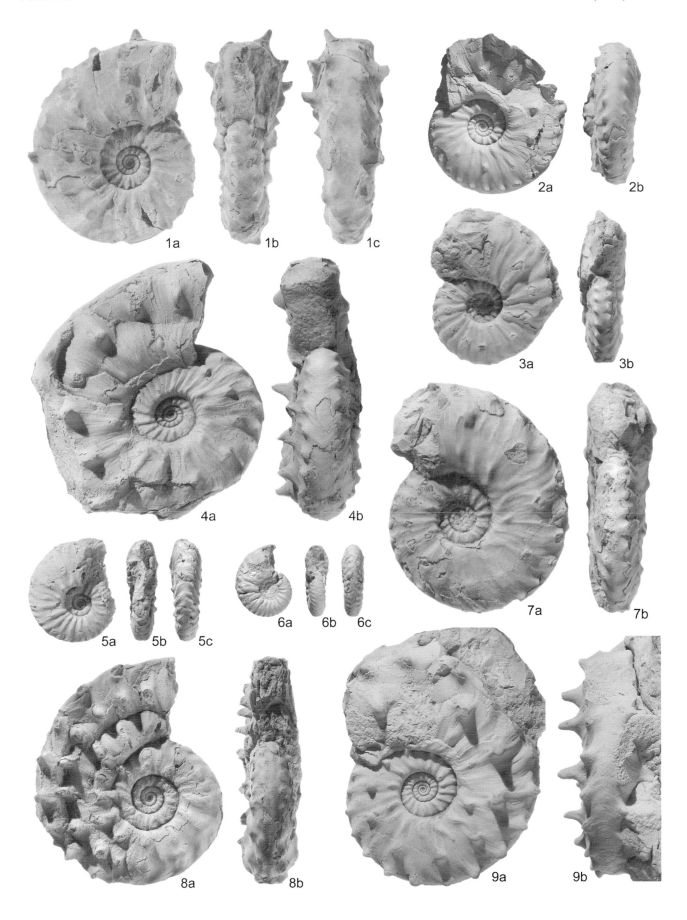

Plate 29 (all figures natural size)

1a–c: ***Brackites vogdesi* (Smith, 1904).** PIMUZ 25193.
    Loc. FHB 15a, Fossil Hill (Humboldt Range). *Vogdesi* Subzone,
    Rotelliformis Zone, Late Anisian.

2a–c: ***Brackites vogdesi* (Smith, 1904).** PIMUZ 25194.
    Loc. FHB 15a, Fossil Hill (Humboldt Range). *Vogdesi* Subzone,
    Rotelliformis Zone, Late Anisian.

3a–c: ***Brackites vogdesi* (Smith, 1904).** PIMUZ 25198.
    Loc. HB 2023, Ferguson Canyon (Augusta Mountains). *Vogdesi*
    Subzone, *Rotelliformis* Zone, Late Anisian.

4a–b: ***Brackites vogdesi* (Smith, 1904).** PIMUZ 25197.
    Loc. HB 2023, Ferguson Canyon (Augusta Mountains). *Vogdesi*
    Subzone, *Rotelliformis* Zone, Late Anisian.

5a–b: ***Brackites vogdesi* (Smith, 1904).** PIMUZ 25195.
    Loc. FHB 15a, Fossil Hill (Humboldt Range). *Vogdesi* Subzone,
    Rotelliformis Zone, Late Anisian.

6a–b: ***Brackites vogdesi* (Smith, 1904).** PIMUZ 25199.
    Loc. HB 742, Oliver Gulch (Augusta Mountains). *Vogdesi*
    Subzone, *Rotelliformis* Zone, Late Anisian.

7a–b: ***Brackites vogdesi* (Smith, 1904).** PIMUZ 25196.
    Loc. FHB 15a, Fossil Hill (Humboldt Range). *Vogdesi* Subzone,
    Rotelliformis Zone, Late Anisian.

8a–b: ***Brackites vogdesi* (Smith, 1904).** PIMUZ 25202.
    Loc. HB 2023, Ferguson Canyon (Augusta Mountains). *Vogdesi*
    Subzone, *Rotelliformis* Zone, Late Anisian.

9a–b: ***Brackites vogdesi* (Smith, 1904).** PIMUZ 25201.
    Loc. HB 2023, Ferguson Canyon (Augusta Mountains). *Vogdesi*
    Subzone, *Rotelliformis* Zone, Late Anisian.

10a–b: ***Brackites vogdesi* (Smith, 1904).** PIMUZ 25200.
    Loc. HB 2023, Ferguson Canyon (Augusta Mountains). *Vogdesi*
    Subzone, *Rotelliformis* Zone, Late Anisian.

11a–d: ***Brackites spinosus* n. gen. n. sp.** PIMUZ 25183, holotype.
    Loc. HB 703, Oliver Gulch (Augusta Mountains). *Vogdesi*
    Subzone, *Rotelliformis* Zone, Late Anisian.

12a–b: ***Brackites spinosus* n. gen. n. sp.** PIMUZ 25182, paratype.
    Loc. HB 703, Oliver Gulch (Augusta Mountains). *Vogdesi*
    Subzone, *Rotelliformis* Zone, Late Anisian.

PLATE 29

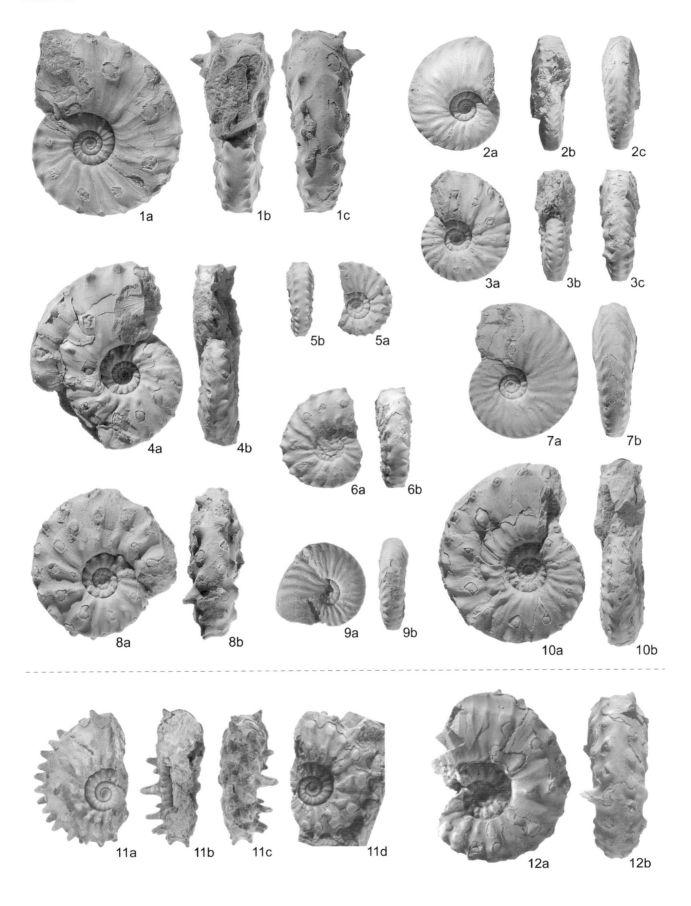

Plate 30 (all figures natural size)

1a–b: ***Eutomoceras dunni* Smith, 1904.** PIMUZ 25258.
Loc. FHB 34, Fossil Hill (Humboldt Range). *Blakei* Subzone, *Rotelliformis* Zone, Late Anisian.

2a–b: ***Eutomoceras dunni* Smith, 1904.** PIMUZ 25259.
Loc. FHB 34, Fossil Hill (Humboldt Range). *Blakei* Subzone, *Rotelliformis* Zone, Late Anisian.

3a–b: ***Eutomoceras dunni* Smith, 1904.** PIMUZ 25260.
Loc. FHB 34, Fossil Hill (Humboldt Range). *Blakei* Subzone, *Rotelliformis* Zone, Late Anisian.

4a–c: ***Eutomoceras dunni* Smith, 1904.** PIMUZ 25261.
Loc. FHB 12, Fossil Hill (Humboldt Range). *Vogdesi* Subzone, *Rotelliformis* Zone, Late Anisian.

5a: ***Eutomoceras* cf. *E. dalli* Smith, 1914.** PIMUZ 25257.
Loc. FHB 31, Fossil Hill (Humboldt Range). *Blakei* Subzone, *Rotelliformis* Zone, Late Anisian.

6a–c: ***Proarcestes* cf. *P. bramantei* (Mojsisovics, 1869).** PIMUZ 25355.
Loc. HB 2062, McCoy Mine (New Pass Range). *Cordeyi* Subzone, *Weitschati* Zone, Late Anisian.

7a–c: ***Proarcestes* cf. *P. bramantei* (Mojsisovics, 1869).** PIMUZ 25356.
Loc. HB 593, Oliver Gulch (Augusta Mountains). *Cordeyi* Subzone, *Weitschati* Zone, Late Anisian.

8a–d: ***Proarcestes* cf. *P. bramantei* (Mojsisovics, 1869).** PIMUZ 25357.
Loc. HB 593, Oliver Gulch (Augusta Mountains). *Cordeyi* Subzone, *Weitschati* Zone, Late Anisian.

9a–b: ***Proarcestes* cf. *P. bramantei* (Mojsisovics, 1869).** PIMUZ 25358.
Loc. FHB 15a, Fossil Hill (Humboldt Range). *Vogdesi* Subzone, *Rotelliformis* Zone, Late Anisian.

10a–b: ***Tropigastrites louderbacki* (Hyatt & Smith, 1905).** PIMUZ 25455.
Loc. FHB 34, Fossil Hill (Humboldt Range). *Blakei* Subzone, *Rotelliformis* Zone, Late Anisian.

11a–b: ***Tropigastrites lahontanus* Smith, 1914.** PIMUZ 25452.
Loc. FHB 34, Fossil Hill (Humboldt Range). *Blakei* Subzone, *Rotelliformis* Zone, Late Anisian.

12a–b: ***Tropigastrites lahontanus* Smith, 1914.** PIMUZ 25454.
Loc. FHB 17, Fossil Hill (Humboldt Range). *Vogdesi* Subzone, *Rotelliformis* Zone, Late Anisian.

13a–b: ***Tropigastrites lahontanus* Smith, 1914.** PIMUZ 25453.
Loc. FHB 30, Fossil Hill (Humboldt Range). *Blakei* Subzone, *Rotelliformis* Zone, Late Anisian.

PLATE 30

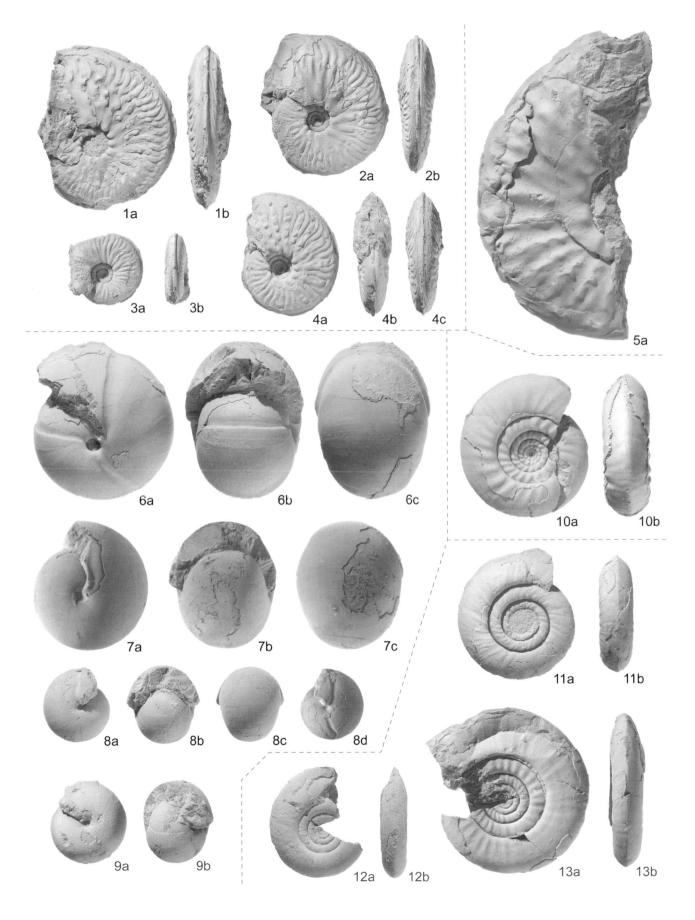

1a   1b   2a   2b   3a   3b   4a   4b   4c   5a

6a   6b   6c   10a   10b

7a   7b   7c   11a   11b

8a   8b   8c   8d

9a   9b   12a   12b   13a   13b

Plate 31 (all figures natural size)

1a–c: *Longobardites parvus* (**Smith, 1914**). PIMUZ 25337.
Loc. HB 590, Oliver Gulch (Augusta Mountains). *Cordeyi* Subzone, *Weitschati* Zone, Late Anisian.

2a–c: *Longobardites parvus* (**Smith, 1914**). PIMUZ 25338.
Loc. HB 584, Oliver Gulch (Augusta Mountains). *Cordeyi* Subzone, *Weitschati* Zone, Late Anisian.

3a–b: *Longobardites parvus* (**Smith, 1914**). PIMUZ 25343.
Loc. HB 740, Ferguson Canyon (Augusta Mountains). *Cordeyi* Subzone, *Weitschati* Zone, Late Anisian.

4a–c: *Longobardites parvus* (**Smith, 1914**). PIMUZ 25339.
Loc. HB 584, Oliver Gulch (Augusta Mountains). *Cordeyi* Subzone, *Weitschati* Zone, Late Anisian.

5a–c: *Longobardites parvus* (**Smith, 1914**). PIMUZ 25341.
Loc. HB 584, Oliver Gulch (Augusta Mountains). *Cordeyi* Subzone, *Weitschati* Zone, Late Anisian.

6a–c: *Longobardites parvus* (**Smith, 1914**). PIMUZ 25342.
Loc. HB 713, Oliver Gulch (Augusta Mountains). *Cordeyi* Subzone, *Weitschati* Zone, Late Anisian.

7a–c: *Longobardites parvus* (**Smith, 1914**). PIMUZ 25340.
Loc. HB 584, Oliver Gulch (Augusta Mountains). *Cordeyi* Subzone, *Weitschati* Zone, Late Anisian.

8a–c: *Oxylongobardites acutus* **n. gen. n. sp.** PIMUZ 25351.
Loc. HB 2030, Ferguson West (Augusta Mountains). *Cordeyi* Subzone, *Weitschati* Zone, Late Anisian.

9a–c: *Oxylongobardites acutus* **n. gen. n. sp.** PIMUZ 25352, holotype.
Loc. HB 596, Oliver Gulch (Augusta Mountains). *Transiformis* Subzone, *Weitschati* Zone, Late Anisian.

10a–c: *Longobardites zsigmondyi* (**Böckh, 1874**). PIMUZ 25346.
Loc. HB 735, Muller Canyon (Augusta Mountains). *Lawsoni* Subzone, *Mimetus* Zone, Late Anisian.

11a–c: *Longobardites zsigmondyi* (**Böckh, 1874**). PIMUZ 25347.
Loc. FHB 9, Fossil Hill (Humboldt Range). *Lawsoni* Subzone, *Mimetus* Zone, Late Anisian.

12a–b: *Longobardites zsigmondyi* (**Böckh, 1874**). PIMUZ 25348.
Loc. FHB 9, Fossil Hill (Humboldt Range). *Lawsoni* Subzone, *Mimetus* Zone, Late Anisian.

13a–c: *Longobardites zsigmondyi* (**Böckh, 1874**). PIMUZ 25345.
Loc. HB 735, Muller Canyon (Augusta Mountains). *Lawsoni* Subzone, *Mimetus* Zone, Late Anisian.

PLATE 31

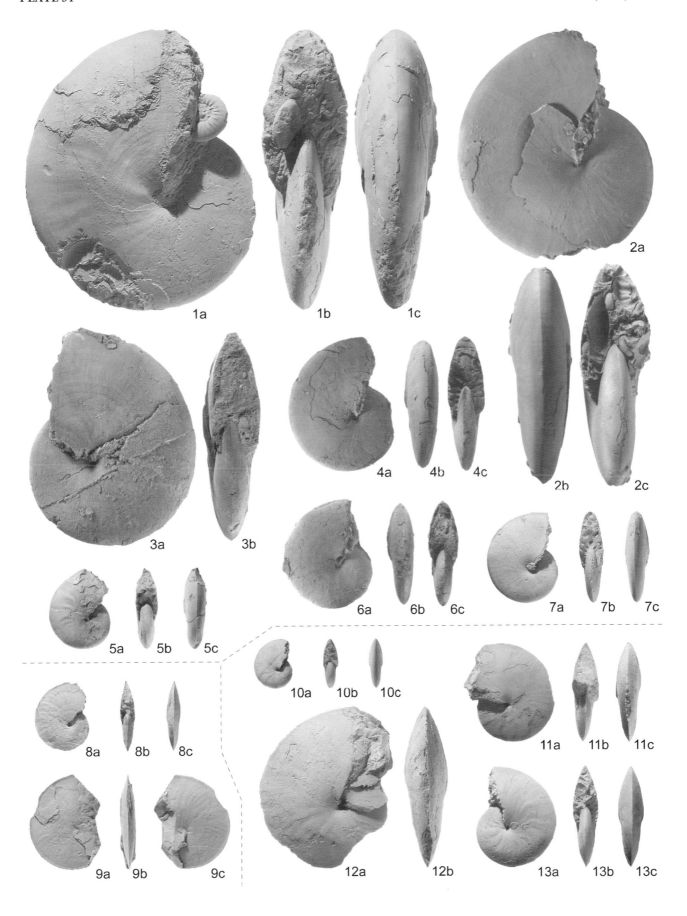